AF302549

FSC
www.fsc.org
MIXTE
Papier issu
de sources
responsables
Paper from
responsible sources
FSC® C105338

Gérald Vignaud

Change the world ?

Understanding the unprecedented
challenges of the 21st century

Foreword

A few months ago, I wrote "School is important but education is paramount !". It is a book for personal development that talks about understanding the world, communication, strategies, health, ecology and building the future. I wrote it instinctively and passionately because it is the personal guide that I would have liked to have had when I was younger. The one I would like to pass on to my son when he is old enough to be able to read and understand it. It is a very complete book of more than 600 pages written in small print. From experience, I know that only a minority of people read books of this size. To make its content accessible to the greatest number of people, I have therefore subdivided it into five small books. Five essential themes that make up the collection "Education is paramount !". "Change the world ?" is one of these five books, a guide that I am happy to share with you today.

I hope that no matter what your age and current life situation, it will bring you some of the keys you are looking for. Also, forgive me in advance for the few "swear words" that you will find here and there in this book. I am not naturally vulgar, but each word has a unique and precise emotional charge, so I have found it useful to use a few of them occasionally, in order to support some of my words even more. Also, as you will notice, this book is written in the masculine form. One must of course see behind this editorial approach the idea of a 100% neutral and generic communication. I am obviously addressing everyone here, girls and boys alike.

As you will discover, I ask many questions in this book to which I invite you to reflect and answer in all sincerity. Also, regarding its use, do not hesitate to break the rules and read it with a pen at hand. Write directly on it your answers to the questions in the exercises. Write in the margins and on the blank pages all the ideas and thoughts that come to you. Highlight the passages and quotations that inspire you in fluorescent, dog-ear the pages and don't be afraid to damage them. Never lose sight of the fact that a stoned book from which you have drawn and integrated all the ideas is ten thousand times more valuable than a book that has never been opened and has been stored for years on a dusty shelf.

Also, I want you to know that your feedback and ideas are essential to me. They help me to constantly question myself and to continuously improve what I have been doing for the past 20 years. In this age of the Internet and horizontal communication, reading a book without being able to communicate with its author seems to me to be, from my point of view, an inconsistency. Let's therefore break this traditional pattern together and give us the possibility to contact each other if necessary. For this, I have set up a free and private contact form on my site at the following address:

www.geraldvignaud.com/livre-contact

Don't hesitate to go there to share your feelings about this work with me. I personally read all the messages and try to answer them as often as possible.

You can also find me on social networks:

As well as on my website:

www.geraldvignaud.com

See you soon,

with respect,

Gérald Vignaud

Contents

It was simply inconceivable for me to write a book like this without evoking this magnificent poem by Rudyard Kipling. He wrote it in 1910, for his then 13-year-old son.

Published under the title "If", this text is, in my opinion, one of the most beautiful and powerful poems ever written. And since it has now fallen into the public domain, it is therefore with immense pleasure that I share it here, as a preamble to this work, to invite you to (re)discover it. ☺

____

If (You will be a man, my son!)

If you can keep your head when all about you
    Are losing theirs and blaming it on you,
If you can trust yourself when all men doubt you,
    But make allowance for their doubting too;
If you can wait and not be tired by waiting,
    Or being lied about, don't deal in lies,
Or being hated, don't give way to hating,
    And yet don't look too good, nor talk too wise:

If you can dream—and not make dreams your master;
    If you can think—and not make thoughts your aim;
If you can meet with Triumph and Disaster
    And treat those two impostors just the same;
If you can bear to hear the truth you've spoken
    Twisted by knaves to make a trap for fools,
Or watch the things you gave your life to, broken,
    And stoop and build 'em up with worn-out tools;

If you can make one heap of all your winnings
    And risk it on one turn of pitch-and-toss,
And lose, and start again at your beginnings
    And never breathe a word about your loss;
If you can force your heart and nerve and sinew
    To serve your turn long after they are gone,
And so hold on when there is nothing in you
    Except the Will which says to them: 'Hold on!'

If you can talk with crowds and keep your virtue,
    Or walk with Kings—nor lose the common touch,
If neither foes nor loving friends can hurt you,
    If all men count with you, but none too much;
If you can fill the unforgiving minute
    With sixty seconds' worth of distance run,
Yours is the Earth and everything that's in it,

 And -which is more- you'll be a man, my son!

Rudyard Kipling (1865-1936)

11

<u>Part 1</u>

—

Loving, Growing and Giving,
the three ultimate goals of all life!

« Life is a mystery to be lived, not a problem to be solved. »

Gandhi

The film ''ZEITGEIST: Moving Forward'', the third part of Peter Joseph's exceptional trilogy[1] (which I strongly invite you to discover), begins with this story by Jacque Fresco. I quote him:

My grandmother was a wonderful woman, she taught me how to play Monopoly. She understood that the aim of the game is to acquire and that by accumulating all she could, she would become the "mistress of the game". Then she always told me the same thing: one day you too will learn to excel in this game.

One summer, I played Monopoly almost every day, all day long. And that summer, I learned to master all facets of the game. I came to understand that the only way to win is to devote yourself totally to acquiring, that money and possession are the means to score points. And by the end of the summer, I had become even more ruthless than my grandmother. I was even willing, if necessary, to bend the rules to win the game.

In the fall, we played together again. I took everything she had from her. I watched her give her last dollar and quit the game, completely beaten. That's when she taught me something more. She just said to me:

- Now everything goes back in the box! All those houses and hotels. All the railways and utilities. All these goods and all this wonderful money. Now it all goes back in the box because none of it really belonged to you. You were excited about all these things for a while, but it was there long before you sat down at that table and it's going to be there long after you're gone: players come, players go. Houses and cars, titles and clothes, even your body. Everything you grab, consume and hoard in your life will go back in the box at the end and you will lose everything.

So, you have to ask yourself: What will happen when you finally get the ultimate promotion? When you've made the ultimate purchase? When you've bought the house of your dreams? When you've secured your savings and climbed the ladder to the highest level of success you can achieve? When the passion will fade away - because it will fade away - ? What will happen next? How far do you have to go on this road before you see where it takes you? Obviously, one day you will understand that it will never be enough. So, today you have to ask yourself the question:

"What is really important? »

« The meaning of life is what is left when you get rid
of everything that is absurd. »

Juli Zeh

Indeed, if we take a step back and a minimum of objectivity, it seems indisputable that Jacque Fresco is right when he explains that at the end of the game, everything goes back in the box. So, since we are only passing through, invited for a tiny spark of time on - as Voltaire so rightly said - an atom of mud, what would be the ultimate meaning of life?

For my part, I believe that the fundamental meaning of life can be summed up in three simple but absolutely essential things: **Loving**, **Growing** and **Giving**.

To be convinced of this, you only have to look at a 2/3 year old child, during his very first years, those in which his Mummy -derived by the oxytocin generated by his birth- preserves him a maximum from the negative of our world and offers him a lot of love. During this very special period of life when he is not yet subject to massive social conditioning, the only objectives of a very young child's day are simple: To Love, Grow and Give.

Loving

Loving is something we all experience more or less in our lives. If we go a little deeper into the subject, we realize that there are four different levels of love:

1/ **Self-centred** love: I receive but I don't want to give anything. I demand love!

2/ Love **with compensation**: I give you love but I expect something in return. (To put it plainly, I'm being a whore.)

3/ **True** love: I give love because I want to give it, without asking for anything in return.

4/ **Unconditional** love: I love everyone, including those who don't love me and those who have hurt me.

It is this fourth level of love that we are talking about here -an unconditional love that we will offer to everyone, regardless of who they are and what they have potentially done to us - that we must aim for. For many, it's a really hard level to reach, and only a tiny minority of people achieve it. And these people have a very great power because love is the most powerful force in the Universe. A force that is both colossal and fundamental and that allows the development of an extraordinary will and determination. Love magnifies all experiences and has the power to unite, guide and liberate beings. Love is a force for forgiveness, healing and healing. Love is the only way to grasp the true depth of someone, to penetrate the very essence of their personality. It allows the lover to discover, beyond appearances, the essential traits of the loved one and thus open up a whole world of new possibilities.

To be successful in life is indisputably to love oneself, to love others and to love life. All forms of life. And to love, nothing could be simpler. When you are with someone -whoever you are and whatever the environment and situation- simply, sincerely and secretly think "I love you" against them. Feel love and gratitude for that person in particular and for life in general. Feel gratitude for the Universe and thank it inwardly for all the things it has already given you and for all the things it is preparing to give you. If I am sure of only one thing, it is that it is unquestionably through love and gratitude that the human being will find salvation.

« There is an efficiency inspired by love which goes far beyond and is much greater than the efficiency of ambition; and without love, which brings an integrated understanding of life, efficiency breeds ruthlessness. »

Jiddu Krishnamurti

Growing

In recent decades, personal development has become one of the dominant religions in our increasingly individualized societies. And the industry that has developed behind it generates huge amounts of money in books, courses, conferences and seminars. But while more and more people want to grow -and that's a very good thing- many of them dream of finding miracle recipes. A large part of this famous industry that I have just mentioned, sometimes very expensive and always with pleasure, sells them. Tricks that are supposed to make them evolve quickly while obviously avoiding effort and suffering. The objective of these people is often the same: to become more in order to be able to earn and own more.

From my point of view, I believe that personal development is first and foremost about getting to know yourself. Analyzing your "source code" -why and how you react to events- and understanding your personal functioning.

Then, I believe it is to understand the Universe in which we live. The functioning of its elementary bases such as mathematics, physics, biology, botany, geography, astrophysics ...etc, is the key to understanding the Universe. To become aware of the infinitely large as well as the infinitely small. To realize that we are only invited to this planet for a tiny spark of time.

To grow is also, since on the one hand we live in it and on the other hand we owe it to ourselves to contribute to making it better, to understand the human world. The functioning of all these things, at the origin of abstract and completely artificial ideas, which today have become the elementary pillars of our societies: money, the concept of growth, the media, advertising, mass entertainment, multinationals, freemasonry, lobbying, GAFA, BATX, capitalism, debt, tax havens, the stock exchange, High Frequency Trading (HFT), the world economic crisis, crypto-money, etc. All these things, which did not exist on Earth during the almost 4.5 billion years before the advent of Homo sapiens, have now become very concrete. Forces that have seized control of the direction of our planet's evolution to such an extent that they have a right to life or destruction over things that were originally very real - and above all indispensable - such as nature and animals.

All this obviously leads us, as a species, to question ourselves and try to understand three very important things.

➢ Where do we come from?

> o From the birth of the Universe, followed by the birth of our solar system, the appearance of life on Earth, the advent of the first hominids, the emergence of Homo sapiens, the beginning of the Neolithic period, the birth of civilizations and the entire history of the world up to the present day.

➢ Where are we today?

> o The different situations in which our world finds itself today are not the result of chance or destiny. It is a multitude of events combined together that trace a unique trajectory from our past to our present. And this unique trajectory that stops at today leaves an infinite number of possible trajectories in potential that will start from the present to the future. And for this reason, it is only by having answered these first two questions correctly that we can try to answer the third one:

➢ Where are we going?

> o And to this important question, we must understand that the answer is not a destiny written in advance. We will go where, individually and collectively, we decide to go. It is up to us to decide to make the right choices.

Finally -and above all- I believe that growing up means developing, in an endless journey, one's talents, skills, passions, know-how, interpersonal skills, compassion, attitude, communication, ability to resist, ability to let go, sincerity and so much more. Things that will help to lead us towards a spiritual awakening and personal growth whose ultimate goal is to give and contribute.

Giving

A woman walks down the street with her 5-year-old daughter holding two apples, one in each hand. Together, they pass a homeless man sitting on the ground who calls out to them and tells them he is hungry. The mother, in a mixture of embarrassment and empathy, doesn't dare say no to him. Not having any change on her, she turns to her daughter and asks her to give one of her two apples to this hungry man. The little girl then had a surprising gesture: she quickly brought the first apple to his lips and then bit a piece of it. Even before her mother could react, she did exactly the same with the second one. Shocked, the mother raised her voice:

- How dare you do that, my daughter, when this gentleman is sleeping outside and he is hungry?

Without listening or responding to her mother, the little girl continued her movement, handing one of the two apples to the man sitting on the floor. She smiles at him and says:

- Take this one, it's the best of both! ☺

This story shows us two things:

- o   The first is that we adults are inundated with inopportune programming that too often prevents us from no longer believing in innocence itself.

- o   The second is that the essence of a child's programming is sharing. It is only we adults who corrupt from our beliefs this natural programming that we cause children to lose as they grow older. Perhaps we should then give all the children of the world this wise advice that I once read on the T-shirt of a person I met on the street: "Hey, child, don't grow up, it's a trap! »

Perhaps we should also learn to go back to childhood and reconquer this natural notion of sharing. Sharing, an element that will be indispensable in transforming our sick world.

« Why always sell when there is so much to give? »

Jean-Jacques Goldman

I deeply believe that sharing, giving and contributing are among the essential goals of life. When it is done sincerely and freely, without expecting anything in return, not even recognition, it surpasses one's own person and raises one's life to a higher level. There is spirituality in sharing and contributing.

Redefining success

Nowadays, it would seem that the vast majority of Homo sapiens populations have become disconnected from these three essential fundamentals of life. It is as if the very essence of Loving, Growing and Giving had been completely crushed by the complexity of our current societies engaged in a crazy race. A world that, in the exact opposite of these values, has erected a predominant model of success based on money, ego, fame, power, outward beauty, recognition, immediate pleasure and diversity - not quality and depth - of life experiences.

A model amplified by the advent of social networks that have pushed the vast majority of people, especially in the younger generations, to put themselves on stage. Nowadays, a huge part of the population competes in ingenuity to make the world believe in the truth of their supposed success when, paradoxically, most of them are very far from it. To such an extent that there is now an almost un-written law that is almost unimaginable and which is verified in more than 99% of cases: "The more you seem to have an extraordinary life on the social networks, the more you have a shitty life in reality…".

And this growing discrepancy between the image we want to show and the factual reality we experience is encouraged by the permanent overbidding of the image given by "the others" who are also voluntary prisoners of the same stupid process. All this drastically increases the feeling of emptiness and loneliness, thus accentuating, in an endless vicious circle, the gap between the image one is trying to give and the factual reality of one's life...

And besides, how many of these people with "such an extraordinary life" are empty and depressed when they go to bed at night, when they find themselves alone with themselves? How many of these "thousands of friends following them on social networks" have trouble finding someone to water their plants when they go on holiday?

And while we're on the subject, do you know people who are in this situation? Maybe even among these people who are so imprisoned by their appearances there is someone you know really well? Do you know who I'm talking about?

« The biggest mistake of our time has been to bend human's spirit toward the search of material goods. We have to lift up human's spirit, to bring it to consciousness, beauty, justice, truth, altruism. There only you will find peace with mankind and with it, society. »

Victor Hugo

As the Dalai Lama quite rightly remarked, the planet does not need more successful people (when he says "successful people", he is obviously referring to the predominant model of success built by our societies that we have just discussed, and not the one we will discuss below). On the contrary, he said, the planet desperately needs more peacemakers, healers, restorers, storytellers, and lovers of all kinds.

What if, instead of being a slave to this predominant model that society imposes on us, you redefined your own model of success? A redefinition that could involve a liberation from material things and more spiritual awakening. Through a personal capacity to refocus on your inner self? To free yourself from your addictions? To experience serenity and inner peace? To be happy and to share this happiness around you? To contribute to a better world? Because the world, at this very special time in the history of mankind, desperately needs you?

What is my (new) personal definition of success?

____

____

____

____

____

____

____

____

____

____

____

____

Success according to Ralph Waldo Emerson (1803-1882): To laugh often and much; to win the respect of the intelligent people and the affection of children; to earn the appreciation of honest critics and endure the betrayal of false friends; to appreciate beauty; to find the beauty in others; to leave the world a bit better whether by a healthy child, a garden patch, or a redeemed social condition; to know that one life has breathed easier because you lived here. This is to have succeeded.

A very special moment in the history of mankind

The long history of the world gives birth to a particularly complicated and delicate beginning of the 21st century. Before our very eyes, our era is witnessing the emergence of a multitude of new factors that are colliding. They are radically transforming our planet and our civilizations, faster and faster and not always for the better. We can mention the following in particular:

- <u>An unprecedented demographic explosion</u>

In 2020, there are already more than 7,7 billion (7 700 000 000) Homo sapiens on Earth, and the demographic explosion that began 150 years ago is becoming a little more pronounced every day. We will come back to this in more detail in the next part of the book but, to sum it up in a few words, on our planet, 160 000 people die every day for every 400 000 born.[2] This means that every day the planet is populated by 240 000 more Homo sapiens. With, of course, all that this implies in terms of pressure on resources, starting with water and food.

- <u>The acceleration of the financialization of our economies</u>

One of the major and unexpected consequences of the 2008 financial crisis is the acceleration of the financialization of our economies. More than ever before in our history, capital now has much more value than labor. Add to this the emergence and explosion of high-frequency trading, and the stage is set for much more serious financial crises with dramatic consequences.

« Finance is a weapon of mass destruction. »

- <u>A world that is becoming more and more difficult, especially for women and children.</u>

Taking previous centuries as a point of comparison, humanity in the 21st century seems to have really limited the ravages of wars, epidemics and famines, the three greatest causes of mass death in its history[3]. However, it is facing new challenges that are likely to prove much more complex. For it is a fact that our planet is home to more and more people, has fewer and fewer available resources, and its ecosystems are increasingly damaged. On top of this, the ultra-liberalization of the world is widening the gap between the richest and the poorest more than ever, destroying more and more middle-class people in the process, effectively demoting them to the lower classes.

A world that is becoming more and more difficult and which more often than not pushes everyone towards egocentrism, survival and individualism. And this new world that is being born before our eyes will be particularly complicated for the most vulnerable people, starting with women and children in developing countries. As a reminder, today, 45 million people on Earth live a life of slavery[4], about the population of Spain. And this figure is set to increase over time.

Important accuracy: By the word "slave", I am not talking about people who are voluntarily left locked in a limiting belief system and who work like hell in a shitty job. I'm not talking about the people who earn a miserable wage that is barely enough for them to survive miserably, and there are billions of them. No, I'm talking about "real" slaves -domestic, labor or sexual- in the literal sense of the word.

- <u>An increasingly urbanised world</u>

From the beginning of the Neolithic period, men began to come together to form the first communities, the group promoting work and security. It didn't take long for these first communities to organize themselves into villages. Villages that became towns. It was around the middle of the first millennium BC that the 100 000 population mark seemed to be reached.

However, it was not until the beginning of the 19th century that the million mark was clearly crossed, simultaneously by London and Beijing. Other Western cities such as Paris and New York followed suit. The industrial boom then accelerated this urban growth in Europe, Japan and North America. The shift towards gigantism dates from the second half of the 20th century. Cities with more than 10 million inhabitants - known as megacities- multiplied, mainly in developing countries in Asia, Africa and South America. There are now several dozen of them, London and Paris being far from being the largest.

While they are often out of work and live in villages without access to energy and medical facilities, for many villagers, cities are synonymous with immense opportunities. Attracted by the hope of professional opportunities and a better future, all these populations are therefore drawn to urban centers that are increasingly oversized and polluted. A phenomenon of rural exodus that is accelerating more and more in countries with strong economic development. Every year, millions of people leave the countryside to go to the cities in the hope of a better life. Whereas a century ago, only 10% of the world's population lived in cities, today it is the urban world that reigns with more than half of humanity living in cities. This figure is growing every day and is expected to exceed 2/3 of the world's population by 2050. In the medium term, an additional 2.5 billion people are expected to live in cities, mainly in Africa and Asia, especially in countries such as India, China and Nigeria. This urban population growth will lead to an explosion in the demand for raw materials and energy, and involves huge challenges -pollution, energy, security, transport, housing, lifestyle, etc.- which will have to be resolved.

Most estimates predict that by the end of the century, 75% of the world's population will be urban dwellers.

- <u>An unprecedented evolution of our food</u>

Our diet has evolved more in the last fifty years than in the rest of human history. The health, economic, ecological and ethical challenges associated with these changes are both enormous and unprecedented. It is one of the biggest changes in our evolution and it impacts absolutely the entire world population, present and future. These evolutions are created and controlled in an ultra-unhealthy way by a handful of multinationals that control one of the most important strategic industries there is: our food!

Today's intensive agriculture is industrialized to the maximum, and is being fed with patented seeds, GMOs and pesticides. It is focused solely on maximum yield with no regard for health, crop diversity and respect for soils and ecosystems. On this subject (let's make a small marketing aside), have you noticed that the food industry has managed to reverse the definitions of the products it offers without anyone even noticing? Today, when we speak -for example- of a "normal tomato", it is, by default in the minds of almost all people, a product of intensive agriculture using GMOs and pesticides. A use of genetically modified seeds and chemicals which, according to the elementary laws of Nature, is anything but normal. Conversely, when it is produced without GMOs or pesticides (normally, therefore), the term "tomato" alone is not enough to explain it and it is therefore essential to specify that it is "organic". **In other words, instead of saying "tomato from intensive**

agriculture" and "tomato", we have been conditioned to say and think "tomato" and "organic tomato". And it passed without a hint of Vaseline.

As far as breeding is concerned, things are not getting any better. Today we no longer breed cattle, we produce them! A production whose stated objective is maximum yield. This production is carried out with great reinforcements - most often as a preventive measure - of massive and regular injections of antibiotics and without any respect for the animal and the environment. And I am not talking here about the industrialization at all levels of our food, in which we find massive amounts of sugar, salt, processed fats, additives, colorings and preservatives of all kinds.

- <u>The obesity epidemic</u>

And this major change in our diet brings us its share of hidden health problems, starting with obesity. An epidemic of obesity that now affects more people on the planet than famines[5] -yet it is decimating many people. Whether from supermarkets or restaurant chains, the food we eat every day is riddled with chemicals. These additives are absolutely everywhere and are mostly unknown to most people. For example, a single French fry in a fast-food restaurant can contain up to ten different additives.[6]

The consequences of all this chemistry on our health are very serious: increased risk of cardiovascular disease, heart attack, hypertension, gallbladder disease, sleep apnea, cancer of the uterus, breast, prostate, colon, asthma, infertility, diabetes, deficient liver ...etc. In addition to this, there is the psychological impact, a decrease in the quality of the emotions and the quality of life of those who suffer from it. On average, obesity leads to a loss of life expectancy of about 10 years.[6]

A few data about obesity: At the time of writing, in France, 1 in 6 adults is obese and 20% of children are overweight. In this country, obesity kills tens of thousands of people a year. In the USA, it is worse: 1 child in 3 is overweight. 1 in 5 is obese. In the USA alone, there are 1 100 deaths a day linked to obesity. Worldwide, 2 million people die every year from its consequences. Figures that are constantly rising…

- <u>The tens of thousands of chemicals that have flooded our daily lives</u>

One of the discoveries that Homo sapiens also greatly appreciated was chemistry. He has a lot of fun with it and, whether for his domestic needs, those of his industries or those of intensive agriculture, he has artificially created a multitude of different molecules - the total number of which is estimated at several tens of thousands. Molecules that he uses absolutely everywhere and whose collateral health and ecological damage is incalculable. And this is not to mention the "cocktail effects" linked to the multitude of mixtures -voluntary or not- of these molecules with each other.

- <u>The omnipresence of endocrine disruptors</u>

One of the consequences of this invasion of chemistry in our societies is the omnipresence of endocrine disruptors. In recent decades, this invisible chemical soup of incredible diversity has penetrated every aspect of our daily lives until it has completely submerged us. Very often dangerous for our health, it is omnipresent everywhere around us: toothpaste, soap, shower gel, shampoo, deodorant, perfume, cosmetic creams, make-up, detergent, fabric softener, food additives, cleaning products, textiles, ink, paint, disinfectants and other industrial products of all kinds. Even our curtains and hoover contain them…

These chemical pollutants usually have unpronounceable names such as "triphenyl phosphate", "benzophenone" or "resorcinol". They exist because they "improve" a manufacturing or conservation process or bring more texture, taste, comfort or seduction to the product. But their effects on our health are catastrophic: alterations in the quality of sleep, behavior and mood, imbalance of the hormonal system, cancers, infertility, congenital malformations. Not to mention the residues they generate, which cause irreversible damage to the environment?

« The world could have been as simple as the sky and the sea. »

André Malraux

- <u>Antibiotic resistance</u>

Discovered by Alexander Fleming at the beginning of the 20th century, antibiotics are used to fight bacteria. Penicillin, first used in 1928, was the first of them. For several decades, their development made it possible to cure many diseases. Since then, many people's lives have been saved thanks to antibiotics, and their use has become widespread, to the point where they are now used on a massive scale -as a preventive measure- in factory farming and beekeeping. In humans, they are often prescribed and used when they are not necessary, for example in cases of virus infections.

As a result, because it has been - and still is - overused, antibiotic therapy is becoming less and less effective and thousands of people die every year from infections that were easily curable only 10 years ago... Bacteria are developing more and more resistance to antibiotics, so much so that a growing number of scientists are sounding the alarm and announcing the emergence of a major health problem. Some even mention the possibility of a return to the 18th century, to the pre-antibiotic era, when the slightest infected wound could prove fatal. In early 2016, a British report announced that by 2050, antibiotic resistance would cause the deaths of 10 million people worldwide each year if nothing changes.[7] And nothing seems to be changing...

- <u>Electromagnetic pollution</u>

From the mid-90s, electromagnetic waves have been pouring into the world in massive quantities. Mobile phones, relay antennas and -more recently- WIFI terminals have multiplied at an astounding rate almost everywhere on the planet, even though no independent scientific study has proven their innocuousness. On the contrary, the waves would have a harmful effect, particularly on our central nervous system. Numerous experiments, some of which have been carried out on rats (whose DNA is very close to that of humans) have shown that the waves promote the development of cancerous tumors and damage neurons...[8]

- <u>The massive explosion of knowledge and technology</u>

The digitization of the world coupled with the invention and rise of the Internet has transformed humanity into a gigantic collective brain. This has enabled a massive sharing of knowledge, resulting in an unprecedented explosion of technology. And all sectors are involved, starting with connected objects, augmented reality, holograms, big data, algorithms, artificial intelligence, facial and behavioral recognition, virtual reality, robotics, autonomous cars, drones, nanotechnology, biotechnology, computer science, cognitive science, cloning, 3D printing, blockchain and even quantum computing.

But while it is true that the development of sound technology is most likely part of the solution to many of the major challenges of the 21st century, we must stop believing that the technological explosion alone will be the solution. As Idriss Aberkane rightly pointed out, an exponential growth in technology that is not accompanied by an exponential growth in wisdom will lead us unstoppably into a wall. If we look at them objectively, the incredible technological advances that humanity is currently conceiving seem to bring far more dangers than solutions. And create intensely more needs than they fill...

- The emergence of crypto-currencies

Based on the concept of the blockchain, 2008 saw the birth of Bitcoin, the first crypto-money. Since then, a multitude of them have come into existence -in January 2020, there were 2400 of them; in January 2021, more than 8000. Although most crypto-currencies are worth almost nothing, because they are owned by almost nobody, some of them like Bitcoin, Ethereum and Ripple are changing the rules of the global economy. They bypass all the traditional financial systems, opening the door to a multitude of applications, and present themselves as major players in the future of finance.

Then there's Facebook - with its more than two billion users and its more than borderline integrity - which is planning to launch its own, the already controversial Libra. A launch that, if it goes ahead, promises to be a powerful fragmentation bomb that could have a profound impact on the world.

- The invasion of screens from an early age

One of the major consequences of the digitization of the world is the invasion of screens. Whereas for a very long time they simply did not exist (and everyone was doing very well), in just a few decades they have become omnipresent and indispensable.

Computers, smartphones, tablets, GPS, connected watches, billboards, cash dispensers, cash registers, train ticket dispensers, airport check-in kiosks, information screens, control screens, etc. Today, whether at home, in the business world, in institutions, in administrations, in public places, in the street, in vehicles and even at the petrol station when we fill up, the screens accompany us everywhere, at any time of the day or night. Yes, even at night, because nowadays, how many Homo sapiens fall asleep with their TV on or with their eyes glued to their mobile phones or tablets? And, even though there are far fewer of them, how many wake up at night to see if they've received messages on WhatsApp, Insta or Snapchat to possibly reply? This situation may make you smile, but it's important to know that this is what a large proportion of teens do. This poses a very serious health problem because, as you probably already know, sleep is a crucial time for rest and the formation of neurological connections. Something particularly important and active during the period of adolescence.[9]

In this day and age, it is not uncommon to see parents giving their kids a screen while they are waiting for their meals in restaurants. And when asked why, they say it's the only way to channel them. But if that's the case, then how did parents 30 years ago do it?

Although they can bring real added value in some cases, the invasion of screens brings new challenges to manage: addiction, unhealthy excitement, decreased concentration, blue light (which disturbs sleep) and a tendency to an absolute belief in the information delivered on them are the opposite side of an invention that few people realize how much it has changed the world. Generally speaking, the invasion of screens is a major health problem. But the massive exposure of children to screens, especially from an early age, before they are 5 years old, is an absolute disaster. We are still very far from imagining the repercussions it will have on their construction and personal balance as well as on their adult life, but it is highly likely that they will be really harmful. See you in 20 years...

« What we do to our children is inexcusable. Never
in the history of mankind has such an experience of
decerebration been carried out on such a large scale. »

Michel Desmurget

- <u>The invasion of the fictional at the expense of reality</u>

Imagination and laughter being one of the peculiarities of our species, Homo sapiens has always placed entertainment at the heart of his life. Musicians, dancers, storytellers, comedians, magicians and troubadours of all kinds have always punctuated the life of different human civilizations. But in recent decades, he has profoundly changed his approach to entertainment, transforming it into a gigantic industry. A machine for creating particularly well-designed virtual worlds, such as those of Star Wars, Marvel, The Lord of the Rings or Game of Thrones. Complete ecosystems - films, music, derivative products, video games, amusement parks, rumors...etc. - known and consumed by billions of people across the planet. As a result of this massive invasion of ultra-accomplished and completely addictive entertainment, we unfortunately notice that among today's Homo sapiens, more and more have a better knowledge of the unreal and the virtual than of their own Universe. Indeed, while they are becoming fans of it, there are many more who know all the historical and technical details of a fictional universe created to measure by the entertainment industry. To take *Star Wars* as an example, countless fans know absolutely everything about this universe, including a multitude of improbable details such as the exact size of the Death Star, the raw material used to make Han Solo's blaster, the cruising speed of this or that ship, or the details of the childhood of this or that secondary character in the film.[10]

Let's be clear, there's obviously nothing wrong with being a Star Wars or Game of Thrones enthusiast and fan to the point of knowing all the minute details at your fingertips. The problem is that while they have mastered all the elements of their favorite virtual worlds in minute detail, many don't even know the basics of how their own Universe works. The functioning of all these ecosystems in which they evolve and which, for their part, are very real. Indeed, isn't it hallucinating to discover that more than 9% of French people believe "it is possible that the Earth is flat and not round as we have been told since school"? [11] That one quarter of Americans think that it is the Sun that revolves around the Earth[12] or, even worse, that 7% of American adults (which still makes more than 16 million) believe that chocolate milk comes from brown cows?!? [13]

And I'm not talking about understanding the crucial importance of the oceans and tropical forests in regulating and balancing the climate?

- <u>The individualization of man and the artificialization of his relationships</u>

Today, thanks to his smartphone, anyone can connect with any of the billions of people who make up the Facebook community. In the same way, you can download the Tinder application in just a few seconds and "choose" a person to spend the night (or more) just by swiping to the right on the profiles you like. I'm only talking about two of the most famous platforms, but the reality is that the tools to connect to others are countless and the potential connections almost infinite. The advent of smartphones, social networks and these "disposable dating" applications has installed and generalized a strange paradox in which we see everyone connecting virtually with everyone else and, at the same time, being completely alone, our eyes absorbed by this screen that fits in our hand. A screen that swallows time, absorbs attention and sometimes even destroys neurons. This screen of a smartphone whose only vision of loss or oblivion most often causes its owner's immediate panic.

In this context, while there are infinitely more people one can potentially meet than there is time to meet, the general tendency is to reduce considerably the depth of one's relationships as well as the level of presence with each of them. A trap which thus pushes everyone to become more and more artificial, superficial and individualistic in the "management" of their relationships...

- <u>The polarization of our societies</u>

Social networks - including Facebook, Instagram, Twitter, Snapchat, TikTok, YouTube and Pinterest to name just the most famous - are all at the heart of a new and little-known industry: that of the personal data they collect when using their platforms. And in order to develop an outstanding profitability, these companies have developed a very unhealthy and ultra-efficient business model. In order to keep us stuck for as long as possible in front of the content they broadcast, these new economic giants have developed a system of "filter bubbles"[14]. Thanks to powerful algorithms in perpetual evolution, they provide us with endless customized information that corresponds exclusively to what we want to see and hear; reinforcing our beliefs, whatever they may be, a little more each day...

As a result, every user of a social network will tend to develop the regularly maintained and amplified certainty that his views are THE truth. This often leads to the belief that all those who do not think as he does are necessarily wrong. There is no better way to divide societies from within and thus prepare the ground for violent civil wars...

To learn more about this crucial and more than problematic subject, take the time to read Roger McNamee's "Facebook, the disaster foretold", to look at "Behind our smoke screens" on Netflix and to visit thesocialdilemma.com.

- <u>The rise of crime</u>

In our increasingly hard world, money is mechanically worth more and more. This is leading to a massive rise in crime, which often affects the weakest and least well-informed people first and foremost. This rise has been accentuated by the digitalization of the world, which has invaded our daily lives and of which only a tiny percentage of the population has mastered the basics. In his excellent book "Futures crimes", Marc Goodman estimates that the percentage of the world's GDP with criminal sources is set to double in the coming decades...

- <u>The end of privacy</u>

In recent decades, the exponential evolution of technology and our means of communication has given the word "surveillance" a completely new dimension. Facial and behavioral recognition, geolocation of our computers, GPS and mobile phones, automated license plate recognition, credit card payments, messaging and telephone communications, social network activities, connected objects, games and applications on our mobiles, search engine queries, shopping, YouTube video viewing, web browsers and mailboxes - to name but a few - shape the personal profile of each individual. All our behaviors are tracked, identified, saved, analyzed and grouped together. From the most insignificant to the most relevant, all of the mind-boggling amounts of data we produce every day ultimately form an ultra-precise portrait of each and every one of us. From our fleeting emotional state to our intimate character traits, from our lifestyle and consumption habits, our tastes, influences, relationships, political opinions, spiritual aspirations or even our sexual orientations and excitements, nothing is missing!

Since the attacks of September 11, 2001, many laws have been opportunely passed to legalize and increase such intelligence and data collection of all kinds. In order to offer us, they say, more security, some leaders are making decisions that inexorably lead us to a world that is ultra-connected and under massive and permanent surveillance. These colossal mountains of collected and aggregated data are then preserved over time and form a sort of personal digital shadow that grows a little bigger every day and follows us around indefinitely. But perhaps the worst is yet to come: with each major crisis, the governments of all countries take advantage of the shock and fears it generates to impose even more freedom-destroying laws. Notably in France, where the very strong emotion linked to the terrorist attacks of 2015 has allowed the implementation of a temporary state of emergency... which has since become permanent. We therefore look forward to discovering all the laws that will, in the name of protecting our health, emerge following the COVID 19 pandemic! (which, at the time of writing, has just started)

One of the unhealthy and unfortunate consequences of all this is that a child who is born today will grow up and spend his entire life without any privacy or personal intimacy. Throughout his life, all his ideas, experiences, mistakes, thoughts, interests, actions, successes, failures, opinions and emotions will be tracked, analyzed, stored and exchanged by a multitude of people, companies, institutions and government agencies whose very existence he is unaware of. But, as Edward Snowden rightly points out, the very essence of being and personality of every human being cannot be built without privacy and intimacy.[15]

« Arguing that you don't care about the right to privacy because you have nothing to hide is no different than saying you don't care about free speech because you have nothing to say. »

Edward Snowden

- <u>The tilting of the centers of gravity of the different powers that shape the world.</u>

One of the major changes in recent decades has been the shift in the centers of gravity of the various powers shaping the world. From the west, it is tilting towards the east. From the Atlantic coasts, it is shifting towards the Pacific coasts. From the States, it is shifting towards the multinationals. Multinationals, including digital and new technology companies which, in barely 20 years, have shaken up the global economic game, dethroning the all-powerful financial and fossil fuel industries (but which are doing very well anyway, don't worry). Not to mention all those many silent underground communities that, thanks to their international ramifications and the considerable human and financial resources they hold, place their pawns wherever they can.

And in the midst of all these upheavals, we can see Europe ageing and collapsing on its own weight, while, in addition to India and China, countries with great potential are emerging such as Mexico, Brazil, South Africa, Turkey, Thailand, Vietnam and Indonesia.

- <u>The arms race</u>

All these things have notably resulted in an arms race that is taking off again with, more than ever, its fusion with new technologies. While the use of drones is already widespread, rather frightening concepts such as killer robots equipped with facial recognition are now making their appearance...

« You don't prepare for peace by preparing for war,
you prepare for peace by working for peace. »

Jean-Luc Mélenchon

- <u>The rise of artificial intelligence</u>

« By 2035, we will no longer see the difference between a man and a robot ». This frightening prediction is attributed to David Hanson, CEO of Hanson Robotics, one of the brightest lights in the robotics industry. A unique industry that is growing at an ultra-rapid rate and whose volume of knowledge, at the time of writing, is doubling almost every year! (Take a few seconds to reread this last sentence and try to grasp its meaning and all the consequences).

Driven by incredibly complex algorithms, artificial intelligence invades our lives a little more each day. Since the dawn of time, men have been providing services for other men. However, since yesterday, we have been withdrawing money or buying our train ticket from an ATM. Today, a hotel in Nagasaki, Japan, welcomes its guests exclusively with the first generations of intelligent robots.[16] Meanwhile, an artificial intelligence made by Google is winning against Lee Se-Dol, the world champion of the game of Go.[17] With each passing day, AI invades our lives a little more. And tomorrow, all tasks -from the most mundane to the most complex and specific- will be performed by robots capable of interaction and reflection. Tomorrow is barely 10 to 20 years away.

The day after tomorrow, humanoids will be everywhere. And, according to David Hanson, they will look so much like humans that they will be confused with them. (It's up to you to imagine what a world like this could look like...)

Behind these major scientific and technological advances lie considerable and irreversible economic, political, cultural and societal changes. They are accompanied by ethical questions and unprecedented challenges. Our world is not just changing, it is radically transforming. In a world that is already increasingly confusing the rule and the human, what will happen when tomorrow all services including Information, Health, Education, Police and the Army will be provided by robots?

« To succeed in creating artificial intelligence would
be the greatest event in the history of mankind!
…But it could also be the last one. »

Stephen Hawking

- <u>The sequencing of the human genome</u>

Following the "Human Genome Project" programme undertaken in 1988, the 21st century opened with a major scientific event: the sequencing and assembly of the human genome.[18] This explosion of constantly growing genetic knowledge opens the way to numerous technologies and applications. Innovations that defy elementary ethical rules as we have always known and felt them. And among these applications, there is in particular ''the increased human being''.

- <u>The advent of transhumanism</u>

The exponential evolution of the so-called NBIC technologies (N for "Nanotechnologies", B for "Biotechnologies", I for "Information Technologies", C for "Cognitive Sciences") allows us to envisage a transhumanization of our civilizations in the very near future. Transhumanism, this current of thought that has become a lobby, advocates the use of technology to improve health, repair the body, or even increase the physical, mental, emotional and reproductive capacities of the human being.

Hybridization between man and machine is in progress, and all these technologies that are being developed will, in a few generations, overturn absolutely all our relationships with the world. This augmented man, who has always been pure fiction, becomes at the dawn of the 21st century an increasingly present reality. Today, science already allows us to choose the color of our eyes, hair or even the sex of our future baby. Like the dystopian film Matrix, will our babies be born tomorrow in artificial wombs?

- <u>A disconnection between man and Nature</u>

Most religions have imposed the idea that man was destined to dominate animals and Nature. An idea on which all our civilizations have been built, and which has resulted in man gradually becoming disconnected from all the ecosystems from which he has come. For a multitude of reasons mainly based on stories of ego, power and money, he has gradually locked himself into a vicious spiral of destruction of Nature. We can say and think what we like, but the reality is that if we look at the situation factually, the observation is irrefutable: our species has today declared war on Nature and, at the same time, is staging its own extinction. Because of the economic systems it has created and which today manage the world, it is systematically wasting and destroying all the natural resources of our planet. The connection with Nature being the primary element indispensable to its spiritual quest, in this sense, we can say that religion has, paradoxically, undeniably distanced Man from spirituality.

« If we look at the facts of the situation, there is no doubt about it:

Our species has now declared war on Nature and, at the same time, is staging its own extinction. »

Gérald Vignaud

In the continuity of its disconnection with Nature, Homo sapiens has, over the course of time and while it has visibly forgotten that it was fully derived from it, dominated the animal kingdom more and more savagely. Whatever they may be, the animals with whom we share the planet have become both helpless spectators and victims of the madness of our species, which classifies them solely in one of two categories: **useful** or **harmful**.

> If the animal is deemed **useful**, Homo sapiens enslaves it by every means at its disposal, even inventing unimaginable infective procedures for the occasion. If you want an example of what I am talking about, type "cows with porthole"[19] on YouTube. Our species has even taken the abominable to extremes by massively institutionalizing animal production and abuse. On Earth today, nearly 1,000,000,000,000,000 (yes, 1 000 billion, or 12 zeros after 1, or 1 million million) animals are "produced" and/or murdered ultra-violently every year just for the pleasure of our taste buds. For your information, the 1 000 billion can be broken down as follows: about 100 billion land animals and about 900 billion marine animals.

> If the animal is deemed to be a **harmful**, then he exterminates it, pure and simple.

« In relations with animals, most people are Nazis
and for animals it is an eternal Treblinka. »

Isaac Bashevis Singer, 1902-1991.

We can say what we want about it but, to date, Homo sapiens is not able to prove formally that on the scale and in the eyes of the Universe the life of a cow, a chicken or even a dragonfly has less value than its own. Even if animal abuse is not a new phenomenon, as Jean de la Fontaine's fable "The man and the adder" testifies[20], it has become institutionalized, industrialized and deeply atrocified in recent decades. **It is urgent that this should stop!**

- <u>Viruses "coming from nowhere" that emerge by surprise from one day to the next.</u>

For those who until then still doubted it, the emergence and lightning spread of COVID 19 across the planet has definitively proved it: a tiny virus of a few tens of nanometers can completely upset the balance of the world in just a few weeks. A virus that can be created artificially by man (and spread intentionally or not), created naturally and transmitted from an animal to man (with a possible nice mutation in the process) or released from the permafrost ice by global warming... (see page 69).

- <u>A major ecological collapse that is becoming more and more inevitable with each passing day.</u>

It is generally very difficult for a majority of people to realize - and even more so to evaluate - correctly the progress of the ecological decay that our planet is undergoing. As everyone considers the environment in which they grew up as the reference for measuring environmental degradation, the perception is biased in advance. Indeed, how can a child who has never run in the woods realize that the forest is damaged? How can a country dweller who sees a few dozen birds a day in his rural sky imagine that only 200 years ago, his ancestors saw several hundred of them in exactly the same place? How can an inhabitant of Paris, Shanghai or New York visualize that, just a few thousand years ago, on the very site of all those immense steel towers and omnipresent concrete, vast wooded areas populated by animals and free of all chemicals and pollution were rising up? Thousands of years which, let's remember, only correspond to a fraction of a second on the scale of the Earth's age.

And yet, as we shall see in more detail in the next part of the book, the ecological collapse facing our planet is already well underway and will accelerate further and further. We are at the dawn of a new era in which all the rules that have always made the world go round will become obsolete. A new and complex world that is racing ahead at breakneck speed and on which no one seems to be able to hold the reins.

« I wish for you all, each of you, to have your own motive for indignation. This is precious. When something outrages you as I was outraged by Nazism, then people become militant, strong, and involved. One joins the current of history and the great current must become continue thanks to each of us. »

Stéphane Hessel

More than ever, we need, individually and collectively, to find a compass. A guide that gives us the right direction to take. What if this guide was simply to reconnect to the basics of life which are Loving, Growing and Giving?

Let's not forget: like in Monopoly, in life, at the end of the game, everything goes back in the box. Absolutely everything! Our possessions, our social status, our successes, our ego... In the end, the only thing left when we're gone is what we loved, gave and passed on.

If these words resonate with you, what do you think about taking a moment to think about them?

What is really the meaning of life? Beyond my daily life, if I take a step back and look at it with the spiritual dimension it deserves, what is, in my opinion, really important?

What does *Love* really mean to me?

What can I change today in my life to integrate more love into it?

What does *Growing* really mean to me?

____

____

____

____

____

____

____

____

____

____

____

What can I change today in my life to incorporate more personal and spiritual growth?

____

____

____

____

____

____

____

____

____

What does *Give* really mean to me?

____

____

____

____

____

____

____

____

____

____

What can I change in my life today to make more of a contribution?

____

____

____

____

____

____

____

____

____

____

____

And since we are talking about contribution, ask yourself:

Do I like the world we live in?

☐ Yes   ☐ No

Why?

What added value would I like to add?

Would I be willing to consider dedicating my life to something bigger than myself? And if so, what would that be? How could I make a positive impact on the world, either in my community or globally?

____

____

____

____

____

____

____

« Whoever we are, we are only one decision away from doing
something that can change the world! »

Edward Snowden

____

Seeing your life as something much bigger than yourself!

Viktor E. Frankl, the author of "Man's search for Meaning", describes his
experience in the concentration camps in the following words: « What was
really needed was a fundamental change in our attitude toward life. We
had to learn ourselves and, furthermore, we had to teach the despairing
men, that it did not really matter what we expected from life, but rather
what life expected from us. »

« it did not really matter what we expected from life, but rather what life
expected from us. » What if that was both the goal and the solution? I
deeply believe that the true value of a person is not in what they earn
and possess, but in what they contribute and bring to the lives of others
and to the planet. Perhaps one of the goals of life would be to dedicate
it to something greater than oneself?

What do
you do in life ?
I am a seller for a shop.
Oh, I wasn't talking about
what you do for a living...
... but what do you do
for the world ?

Maybe if you feel like you were born into a world that doesn't fit you, maybe it's because you've come to help create a new one?

Today, our ecosystems are in danger and more than ever, our Earth needs to be learned, understood, loved and given for it. And that's good, because it corresponds to the deepest goals of all life, including yours...

As with everyone else, one day death will knock at your door. It may well be a unique, powerful and extraordinary experience. Probably the most intense in all of life. And on that day, the only questions that will really matter will be:

- o   What have I done with my life?

- o   Who and how did I love?

- o   What have I learned and understood?

- o   What did I give? What have I contributed to?

- o   What am I leaving as a legacy to this planet that welcomed me?

« I want love, joy and good humor,
It's not your money that makes me happy,
I want to die my hand on my heart... »

Zaz

<u>Part 2</u>

—

We all have the moral responsibility
to protect and preserve our planet

At the very beginning of the Universe, about 13 500 000 000 (13,5 billion) years ago, there would have been the Big Bang. A mysterious event that caused a sudden and colossal explosion. Considerable energy, unimaginable for a human mind, would have spread through space in such a fraction of time that it would be inconceivable to us. The Big Bang thus gave birth to Light, Space, Time and Matter.

Then, in the 9 000 000 000 000 (9 billion) years that followed and even before the birth of ours, countless other stars appeared. They were born, lived and, for many of them, have already died out. In this multitude of potential worlds, it is highly likely that an unimaginable number of incredibly diverse life forms have emerged. Life forms that have developed, evolved and, for probably the vast majority of them, have already disappeared.

Then, about 4.5 billion years ago, our star, the Sun, was born. It was an extremely commonplace average star, resembling several hundred billion other stars, lost on the periphery of its galaxy, the Milky Way. The Milky Way is also itself a galaxy of disturbing banality, wandering through an immense Universe which, according to the latest scientific estimates, is home to nearly 400 000 000 000 000 000 000 (400 000 billion) stars.

By the way, with a minimum of 100 000 000 000 (100 billion) stars per galaxy out of the 400 000 000 000 000 (400 000 billion) galaxies that would exist, that makes 40 000 000 000 000 000 000 000 000 000 (40 quadrillion[21] or 40 million billion) stars in the Universe, each potentially having one or more planets in orbit around it. Of course, we are only counting here the stars that are still alive, excluding those that are dead and already long since extinct, and whose numbers are most likely much larger. From an objective analysis of these empirical data, one deduction is obvious: the possibility that we are the center of a Universe that was created exclusively for us, Homo sapiens, is statistically absurd. The probability is so infinitesimal that it is most likely equal to zero.[22]

« Man is infinitely big in relation to the infinitely small and infinitely small in relation to the infinitely big, which reduces him almost to zero. »

Vladimir Jankélévitch

Our star and its eight planets, including the Earth, were therefore born 4, 500 000 000 (4.5 billion) years ago. And it was 700 000 000 (700 million) years later that the first forms of life, in the form of single-celled bacteria, appeared on Earth, in the oceans. A very slow and complex evolution, with a multitude of pitfalls and extinctions of all kinds, has seen the emergence of a multitude of life forms, the vast majority of which are now extinct. About 7 000 000 (7 million) years ago, the first Hominids - monkeys standing on two legs - appeared. If there is only one thing to remember about this, it is that between the appearance of the first life forms 3 800 000 000 (3,800 million) years ago and the emergence of the first hominids 7 000 000 (7 million) years ago, 3 793 000 000 (3793 million) years have passed.

Over the last 7 million years, several species of hominids have succeeded one another[23] and, in some cases, even crossed paths before the appearance of our own: Homo sapiens. Our species is thought to have appeared around 300 000 years ago, 3 799 700 000 (3 799 million 700 thousand) years after the appearance of life on Earth.

So far, whether for Homo sapiens or for any of the other animal species that have walked the Earth's soil, everything has been going relatively well. Everyone lived in connection with Nature and its immense variety of ecosystems and, apart from very special cases -such as a meteorite ten kilometers in diameter that strikes the Earth off the Yucatan

Peninsula in present-day Mexico[24]- there was an absolutely perfect balance.

The first major shift in this balance occurred about 12 000 years ago with the advent of the Neolithic period. Homo sapiens settled down and invented agriculture and animal husbandry, thus beginning its domination over Nature and animals. Almost 12 000 years have passed since then. 12 000 years during which he developed civilizations and everything that goes with them: possession, power, war, slavery, religions, art, science, commerce, money, writing, bureaucracy, administrations, law, justice, medicine …etc. In spite of the hazards of wars, epidemics and famines which have littered the history of humanity, it has thus multiplied more or less regularly until it reached 1 billion individuals at the beginning of 1800. While the number of his representatives was beginning to become substantial, his very low impact on his ecosystems nevertheless enabled him to maintain a sufficient balance to continue to integrate perfectly into them.

The second major shift in this balance occurred in the middle of the 19th century, on 27 August 1859 to be precise. A summer day that gave birth to the first oil drilling in the history of mankind! Oil, this highly concentrated and easily transportable energy, was the ultimate trigger that multiplied and accelerated everything exponentially. It profoundly changed the face of the world forever. Very quickly, all industries, without exception, were completely turned upside down by the advent of oil. Thanks to it, Homo sapiens has since experienced a period of demented material abundance, one of the main consequences of which was its demographic explosion:

- o At the beginning of 1800, the 1 billion inhabitants mark was passed.

- o In 1927, the 2 billion.

- o In 1960, there were 3 billion of us. At this point, the process accelerates even more.

- o In 1974, there were already 4 billion of us.

- o   In the 1980s, the figure was over 5 billion.

- o   At the dawn of the 21st century, in 1999, there are 6 billion of us.

- o   In 2011, the threshold of 7 billion inhabitants is crossed.

- o   In 2020, the population will be around 7,7 billion people. In less than 10 years, that means 700 000 000 (700 million) more people: a little more than twice the population of the United States!

- o   With a population that is currently growing by **80 million people each year**, projections estimate a population of 8 billion people in 2025, rising to between 9,5 and 10 billion around 2050.

- o   Finally, for 2100, the world population should, according to the various scenarios envisaged -except for that of a major ecological collapse- be between 10 and 25 billion inhabitants.

As already mentioned above, the main reason for this unprecedented growth is very simple: the advent of the economic and health progress brought by oil and the growth in the volume of knowledge. The direct consequences of this have been an increase in life expectancy and, above all, a significant drop in infant mortality, which mechanically increases the number of people of childbearing age. And this exponential population explosion multiplies our impact on our planet's ecosystems.

« Pollution poses new questions that we must absolutely
face up to. It is about life on earth and it is the duty
of all of us, the human species, to preserve it. »

Albert Jacquard

The 21st century, a very special moment in our history

While humanity consumes more than 95 million barrels of oil every day[25] (for information, one barrel is equivalent to 159 liters), oil has become the cornerstone of all global geopolitics. Transport, industry, agriculture, petrochemicals, etc., are all dependent on it. Whether directly or indirectly, oil is hidden absolutely everywhere in our daily lives and all our civilizations have become addicted to it. An oil whose advent has definitively tipped our planet into a major ecological imbalance. An imbalance which, multiplied by our exponential demographic explosion, is accelerating a little more each day.

And finally, in order to further complicate the situation of this world that has become out of control, a third major shift emerged recently, towards the end of the 20th century. A duet, to be more precise, two twin brothers moving hand in hand: the invention (and rise) of the Internet and the digitization of the world.

Among the positive effects of the invention of the Internet and the digitization of the world is the birth of a global collective brain generating the emergence of an explosion of knowledge and technology. The scientific and industrial world is in turmoil and major breakthroughs are being made every day in fields as diverse and varied as genetics, medicine, biotechnology, cloning, cognitive sciences, nanotechnology, information technology, the cloud, connected objects, holograms, augmented reality, virtual reality, artificial intelligence, voice, facial, physiognomic and behavioral recognition, robotics, autonomous cars, drones, and 3D printing.

But alas, since every medal always has two sides, an exponential explosion of technology, if not accompanied by the wisdom that goes with it, will lead us inexorably to our downfall. A lack of wisdom magnificently illustrated by a whole section of these Silicon Valley "visionaries". Like Peter Diamandis, they dismiss the dangers of today's ecological challenges out of hand, explaining that there is no need to worry, that tomorrow's technological discoveries -without specifying which ones- will solve them all.[26] Since a huge part of the decision-making power of the direction the world is currently taking is in their hands, let us hope that their insights are right. But what if they turn out to be wrong?

Is our collective future a danger or an opportunity? As I write these lines, everything is still open, although it seems that the scales are tilting more and more, dangerously and irremediably, towards peril.

The massive exploitation of fossil fuels (oil, gas, coal) as well as an exponential evolution of technologies have been developing for several decades on our fragile little blue planet. Our civilizations are enveloped in a powerful global ideology of infinite economic growth skilfully locked and maintained by very powerful interests. At the beginning of the 21st century, the pressure that we are exerting on the resources of our planet -mathematically multiplied by our unprecedented demographic explosion- is gigantic and, while it should urgently slow down, it tends on the contrary to accelerate more and more.

« Our way of thinking destroys our environment, we would have to change our way of thinking to protect it. »

Steve Lambert

The ecological challenges, both present and future, are many and varied. Today, almost no place on Earth is spared and the challenges related to the preservation of our planet are numerous. It is more urgent than ever to be aware of this and to act. For 160 years, we have been breaking a balance that took 4 500 000 000 (4,5 billion) years to build. But, even if many things are already irreversible, it is not yet too late to react because we can still limit the damage. We must commit ourselves, each at our own level, to do everything we can to repair and preserve what can still be repaired and preserved. We have no other choice because, as things stand, **we are heading straight for a major**

ecological collapse with dramatic consequences. We have a moral responsibility to our children and to all future generations.

The major ecological challenges of the 21st century

At the beginning of the 21st century, Homo sapiens is waking up in a new and increasingly complex world. Completely addicted to oil and digital technology, he must quickly face many ecological challenges. Ecological challenges that are, in one way or another and from near or far, all linked and interconnected. Here are the main ones, with a brief description:

- <u>Climate change</u>

Climate change is an increase in the temperature of the atmosphere caused by the greenhouse effect. When solar radiation reaches us, the Earth reflects some of it back. But when the concentration of greenhouse gases - carbon dioxide (CO2), methane (CH4), nitrous oxide (N2O) and fluorinated gases (HFCs, PFCs, etc.) - increases in the atmosphere, our planet loses its ability to reflect some of this solar radiation, thus mechanically causing a rise in temperature. These few extra degrees completely disrupt the very functioning of our ecosystems and trigger, amplify and/or complicate some of the other major ecological challenges of the 21st century.

Climate change is mainly caused by transport (road, sea and air), all our industries and intensive livestock farming (methane gas released by cattle flatulence). It is particularly accentuated by massive deforestation, acidification of our oceans (which destroys plankton) and the melting of ice (the white reflects and reflects the sun's heat back into space). Human activities, boosted by the mad rush of the market economy and supposedly infinite growth, have brought about this process. Although it has now become impossible to stop it, we must do everything we can to limit it as much as possible. For let us understand the good: **it is a question of survival, nothing less**. But as I write these lines, humanity is unfortunately rather moving in the opposite direction...

On this subject, among all the other crap he tweeted, a certain psychologically disturbed, demagogic and particularly self-centred American president wrote that "If it's a bit warmer? Great, we'll go to the beach more often!". If, like him, you are one of those people who think that if our planet warms up by a few degrees it's not so serious, ask yourself: what if your body temperature rose by 2 degrees, would you enjoy living permanently with a fever of 39°? And if it's not 2 but 4 degrees higher, would you be able to live with a fever of 41° all the time?

I hope that you and your children have all answered "Yes" to these two questions because this is what awaits our planet by the end of the century...

« Instead of wanting to Terraform the planet Mars[27], wouldn't it be much wiser to try not to Venusform the planet Earth?[28]. »

Gérald Vignaud

- <u>Geo-engineering</u>

To solve the climate crisis we are currently experiencing, we have only three options:

- o   Changing our way of life

- o   Adapting to climate change

- o   Climate action

Although it is the most coherent and accessible, the solution of "changing our way of life" involves too many things for too many people to be seriously considered at this point in time.

The second solution, which is to adapt to climate change, will cost -by far- much more than the benefits brought by the industrial and economic growth that gave rise to it.

For some years now, the third hypothesis, intervening on the climate, has been considered and is even beginning to be applied in some cases thanks to a set of techniques aimed at manipulating and artificially modifying it. This is called geoengineering! The advantage of this solution is that it does not go against the powerful lobbies of fossil fuels and does not in any way - in the short term and apparently at least - reduce the way of life of the planet's inhabitants, and therefore of public opinion. It is for this reason that certain political forces are interested in it, supported behind it by people who are betting on the promise of colossal gains and profits. At the Madrid' COP 25, in 2019, the idea took on even greater significance when some people publicly questioned whether it was wise to (attempt to) manipulate the oceans through geo-engineering.[29] Make no mistake, this is only the first step in a very classical communication scheme. You put an idea on the table to start making it familiar. The objective is to make it acceptable, then normal, and finally to make it appear indispensable in the eyes of public opinion in a few years' time. But in the case of geoengineering, the cure is likely to be worse than the disease. Playing sorcerer's apprentice with the climate and the oceans could well disrupt them even more and above all in the long term, with all the dangers that this represents.

When I hear about geoengineering, I can't help thinking of that phrase spoken by Dr. Glen Thompson in the film "Leviathan" (released in 1989). As he was making the observation that the genetic mutations that were tested went wrong, he addressed the other crew members in these words: « **Don't fuck with the Mother Nature!** »

Leviathan (1989) – "Don't Fuck with Mother Nature" Scene : https://www.youtube.com/watch?v=KMl-Sxq9Npg

- <u>The melting of permafrost</u>

One of the consequences of climate change and global warming is the melting of the permafrost. Called permafrost in English, permafrost is the term used to describe all the permanently frozen ground - located mainly in Russia, Alaska and Canada - that covers nearly 25% of the land in the northern hemisphere. As it melts, permafrost threatens to release colossal quantities of C02 currently trapped, estimated at almost 2 000 billion tons (or 2 000 000 000 000 kilos). This gradual release of greenhouse gases will thus accelerate global warming, which in turn will accelerate the melting of the permafrost and so on, creating a dramatic vicious circle. But melting permafrost brings another equally dangerous and unpredictable concern: it could eventually release forgotten viruses and bacteria, which have been hibernating secretly for millennia in these icy soils. Viruses and bacteria against which we would have absolutely no defense...

Few people talk about it, but the melting of the permafrost has already begun. If it continues, it could be one of the worst climate and health time bombs ever to fall on us.

- <u>The destruction of biodiversity</u>

Biodiversity is the sum total of all the animal and plant species and ecosystems evolving on Earth. It ranges from elephants to mosquitoes, from trees to mushrooms, from algae to large marine mammals, including birds, krill and coral. Humans are also part of it.

However, since the 19th century and the exponential and continuous acceleration of its impact on the world, mankind has been dangerously disrupting biodiversity as a whole. Chemistry of all kinds - starting with pesticides - with which we have flooded our planet, the destruction of habitats, "legal" hunting, poaching and global warming are the main reasons for the destruction of biodiversity.

In the last 40 years, we have lost almost half of the living and all the lights are red for a large part of the other half that remains.[30] For example, as far as large mammals are concerned (such as elephants, lions, rhinos, giraffes, gorillas, tigers, jaguars ... etc.), do you realize that at the current rate of extinction, within 2 to 3 decades, that is to say tomorrow, there will only be some left in zoos and children's books? :'(

- <u>The pillage of the oceans</u>

Not to mention illegal fishing -estimated at more than 10 million tons- more than 100 million tons of fish are taken from the world's oceans every year. **That's about 900 billion fishes.** And the demand is constantly growing for fish stocks that are becoming less and less replenished...

Beyond the economic challenge that fishermen will face, this predicted fish shortage will be above all an unprecedented ecological disaster! With current fishing methods and rhythms, in 2048 -**in less than 30 years**- there will be no more fish in the oceans. Literally. Our children will then ask us where the fish we see in the documentaries have gone. We will have no choice but to answer them the sad truth: « We ate them all! »

« Overfishing is the greatest threat to our oceans. »

Lamya Essemlali

- <u>The continents of plastic</u>

Of all the challenges facing humanity in the 21st century, one of the biggest of all will come from plastics. Many people don't know it, but every year more than 10% of the **100 million tons of plastic produced by humanity,** or 10 million tons a year, ends up in the oceans. They form gigantic sheets of plastic soup. In reference to their gigantic size, which is equivalent to several countries, they are called "continents of plastic".

And whether direct or indirect, the impacts of plastic waste are disastrous:

- Unbearable visual pollution on the world's beaches, knowing that what is seen is only a tiny fraction of the waste, most of it being underwater and/or in the middle of the oceans.

- Countless marine animals (turtles, dolphins, sperm whales... etc.) die suffocated by the plastic waste they confuse with their food.

- These plastic plates serve as a support for the implantation of new invasive species, travelling across the oceans and disrupting certain ecosystems.

- Plastic does not biodegrade but micro-fragments. By breaking down into an infinite number of microscopic pieces, it is ingested by fish and enters the food chain.

- Not to mention the seabirds living on these islands thousands of kilometers away from the continents. On their beaches they collect plastic waste brought back by ocean currents and thus die, dying with their bellies full of plastic, guilty of having confused it with their usual food.

And besides, since we are talking about this subject, know that you can, at your own level, act on it today through your daily personal behavior. This can be summed up in six words: Refuse, Reduce, Reuse, Repair, Recycle and Collect. Six "R" words and two "I" words: Inform and Inspire, by example, others to do the same.

Still in a personal capacity, you can also support the politicians who will (one day?) have the courage to stand up against the powerful lobbies of this industry by advocating a significant taxation of plastics. Taxation that will not only make plastics much more expensive to produce - which will mechanically reduce their production and consumption - but will also release funds to subsidise the design, promotion and dissemination of a whole range of ecological alternatives, many of which already exist.

« Plastic floods, disfigures and intoxicates our planet. »

Gerald Vignaud

- <u>Pollution from the shipping industry</u>

Clothing, electronic equipment, DIY materials, toys, furniture, industrial equipment, food, raw materials, vehicle parts, recycled materials, etc. 90 to 95% of everything - absolutely everything - that we consume directly or indirectly passes through the oceans. The international shipping industry thus plays a vital role in our globalized economy. Every day, more than 60 000 container ships -the largest of which can carry more than 23 000 containers[31] - ply the oceans. What is less well known is the numerous collateral environmental damages that this silent industry spreads around the oceans:

- o Residual fuel used for engines is the cheapest but dirtiest fuel available. Extremely polluting, it has a very high content of fine particles, particularly sulphur (it is estimated that a single boat emits as much sulphur as 50 million cars![32])

- o Displacement of invasive species caused by deballasting[33]

- o The noise of engines that deafen countless cetaceans that, disoriented, come ashore and die on the beaches.

- o Wild degassing on the high seas

- o Oil spills

- o The numerous shipwrecks and accidents (on average 1 every 3 days) and the numerous pollution incidents that accompany them.

- o Unrecycled wrecks that end up on rubbish beaches in countries such as India or Bangladesh where the rest of their remains continue to pollute indefinitely...[34]

But with growth set to triple in the next 20 years and the opening of new trade routes to the North made possible by global warming[35], the worst is yet to come and the ecological disaster has only just begun.

« If the Ocean dies, we all die. It's as simple as that! »

Paul Watson

- <u>The extermination of sharks</u>

Although they have been present in the oceans for nearly 450 million years, sharks are now on the verge of extinction. The main cause is the unreasonable appetite of the Chinese for their fins cooked in soup. Despite the exorbitant price and tasteless taste, one legend attributes the strength and vitality of shark fin soup to those who drink it. This makes it a particularly popular dish among China's social elite. A social elite whose numbers have been growing steadily since China's economic boom. The shark fishing industry has become very lucrative - a market worth more than a billion US dollars a year - and exterminates nearly 100 million sharks every year in particularly atrocious conditions. (If you have a weak heart, don't post "Shark finning" on YouTube.) With a reproductive maturity of up to 25 years for some species, you don't need to be a math genius to understand that their extinction is near...

While unknown to the general public, the problem of shark extinction may well be one of the biggest time bombs threatening the oceans. As sharks are top predators at the very top of the food chain, their extinction, through a domino effect, will unpredictably disrupt all our ecosystems.[36]

- <u>The disappearance of bees</u>

Although bees have been present on Earth for much longer than mankind, they are now facing a possible beginning of extinction. For some time now, beekeepers around the world have been observing the disappearance of large numbers of bees in their colonies with increasing frequency. They have called this phenomenon "bee colony collapse syndrome".

Appearing since the 1990s, this worrying phenomenon has grown since the 2000s. Although it is not yet fully understood, scientists and beekeepers have nonetheless established that it is the result of a combination of several factors: the recent proliferation of Asian hornets, the proliferation of parasitic agents -notably the Varroa Destrutor mite-, intensive agriculture and the decline in biodiversity that affect pollen resources and, of course, neonicotinoids, which are massively present in pesticides.

- <u>Deforestation</u>

In the past, men and trees were friends and lived in osmosis together. Today, although forests are essential for biodiversity and climate regulation, Homo sapiens deforests the equivalent of a quarter of the surface area of France every year, **on average the size of a football pitch that disappears every 2 seconds**. The main forests that are victims are the primary forests -also called virgin forests- located in South America, Indonesia and the Congo Basin.

In the Amazon, the forest is razed to intensively raise cattle, which are fed with soya grown on plots of land in other ancient Amazonian forests that have also been razed for the occasion. In Indonesia, the forest is being destroyed for the production of the ever-increasing palm oil. Palm oil is transformed into vegetable oil and is found in products such as shampoos or detergents but also massively in industrial food: pizzas, margarines, biscuits, spreads, ready meals, chips...etc.

Of course, we all agree that we are against deforestation. But if we analyze the situation objectively, in the end it is indeed our lifestyles and consumption patterns that are triggering this massive deforestation...

« The world's forests are home to more than 50% of the planet's biodiversity. This remains largely unknown. »

Yann Arthus-Bertrand

- <u>Intensive breeding</u>

Homo sapiens consumes meat. In ancestral times, he hunted with stones and wooden spears, running behind his prey. In this fight on almost equal terms, the animal was given a chance of survival. Today, Homo sapiens no longer hunts it but massively and violently, obviously delegating this thankless task to others. Every year, about 100,000,000,000,000 (100 billion) land animals are bred -the vast majority industrially in atrocious conditions- to be eaten. And, beyond the abominable animal suffering associated with it, intensive animal husbandry is the cause of much destruction of our resources and environmental degradation. In particular:

- o As mentioned above, the destruction of the forests, especially the Amazon rainforest where the cattle will be "grown" and the destruction of the Amazon rainforest where the soya and maize that will be used to feed the cattle will be grown.

- o Soil destruction through the massive use of pesticides to grow these same soya beans and maize.

- o The colossal and indecent water expenditures that are used to water these cattle and to water the soya and maize that will feed them (it is estimated that to produce just 1 kilo of beef, 13 500 liters of water is needed in the end)[37.]

- o Destruction of soil, river and ocean ecosystems by the colossal quantities of excrement thrown up by all these livestock.

- o And, as already mentioned in the subject of climate change, the flatulence of these billions of cattle that release into the atmosphere phenomenal quantities of methane, a greenhouse gas 25 times more powerful than CO_2.[38]

On this subject, know that there is a very simple way to have an immediate positive ecological impact on the world: **limit or even completely eliminate your consumption of animals!** In addition to no longer participating in the abject animal suffering that humanity has skilfully institutionalized, becoming a vegetarian is today the most important and direct change you can immediately make to help save the planet and its species.

A little imaginary conversation...

- Why do you eat meat?
- Because it is normal, man has always eaten meat!
- Exactly! Like there was a time in your life when you used to poop in your pants. But now you don't do it anymore because you've grown and evolved. Even if it's on a longer time scale, doesn't it make sense for human beings to grow and evolve?

- <u>The invasion of pesticides</u>

At the end of the Second World War, a devastated world was faced with the imperative need to urgently feed its population. It then invented and developed intensive agriculture, which very quickly became dominated by pesticides. A new approach to agriculture with dramatic medium and long-term consequences.

Pesticides are chemicals designed to eliminate pests that can prevent crop development. Intensive agriculture, which probably started out with good intentions, has since developed to become driven solely by maximum yield and profit-seeking. Thanks to cleverly conceptualized communication strategies locked by the industry's lobbies, pesticide use has become massive and systematic. Its consequences are numerous and have serious impacts on the environment (extermination of pollinating insects, progressive destruction of soils, infiltration of water tables) but also on human health (reduced fertility and increased number of cancers).

- <u>Scarcity of arable land</u>

One of the consequences of pesticide use mentioned earlier is the scarcity of arable land. Few people realize this, but in a mere handful of soil, literally billions of living things, from bacteria to earthworms to a multitude of tiny insects, ensure its fertility. 1 gram of soil contains an average of 1 billion bacteria, divided into 10 to 100 000 different species, most of which are completely unknown to us.[39] By constantly flooding all this soil with a lethal chemistry, we are also methodically decimating all the living things it contains. Fertile soil is, however, absolutely living soil. If all this life buried in the earth is destroyed, nothing can grow there anymore.

While there are more and more mouths to feed on our planet, there is at the same time less and less arable land to produce food on. What will happen when there is not enough?

- <u>Air pollution</u>

In recent decades, dust, smoke and *smog* levels have increased dramatically around the world. Human activities are to blame. The air we breathe today is contaminated by industrial waste and automobile transport, but also by land application, toxic smoke from chimneys or, as in India, by millions of open fire cookers.

Today, air pollution is a global problem that affects 9 out of 10 urban dwellers and, according to the WHO, kills nearly 7 million people prematurely every year, mainly in Asia. In a country like France, however, it is estimated that at least 48 000 people die each year from air pollution.[40]

- <u>Light pollution</u>

All living beings on the planet are solar beings driven by natural landmarks linked to the seasons and to day and night. Few people realize this, but the massive and often useless[41] night-time illumination that we impose on our planet not only consumes a colossal amount of energy, but also enormously disrupts -or even disrupt - the surrounding ecosystems that are subject to it.

- <u>The multiplication of our waste</u>

Our daily over-consumption of all kinds of products literally generates mountains of waste. A recent study by the World Bank estimates that we produce 8 to 10 kilos per person per day and that this volume is constantly increasing. And that's just rubbish; this figure does not include other wastes, such as those from energy production, chemical manufacturing, manufacturing, electronics, agricultural waste, and the paper used by more than 7.7 billion people.

Whether it is electronic, plastic or chemical, our waste has a profound and exponential impact on the environment and the multiplication of its production is a global problem. On this subject too, it has become urgent for us to react. Let's hope that this will be the case and that our planet will not end up, as in the dystopian cartoon "Wall-E", buried under waste...

- <u>Widespread pollution of our planet</u>

Definition of pollution: Degradation of an ecosystem by the introduction -usually human- of substances, waste or radiation that alter the functioning of this ecosystem to a greater or lesser extent.

Today, almost no place on Earth is free from pollution linked to human activities. Little by little, we are destroying one by little all the pillars and ecosystems of our planet. A planet that we have a moral responsibility to pass on in good condition to our descendants (as well as, for that matter, to the descendants of all other species). All around the globe, millions of "small" pockets of local pollution, when added together, create a serious problem of widespread pollution of our planet. The examples are countless. Here are just a few of them:

- The Ocean (which, to some, seems conceptually infinite, whereas it is far from being the case) which serves as a dustbin for the nuclear industry with its radioactive waste and other crap illegally dumped into the sea?

- These millions of used nets are lost or abandoned by fishing boats all over the world's seas and oceans (For your information, every year 640 000 tons of fishing equipment ends up on the ocean floor). Quite rightly called "ghost nets", they drift indefinitely and massacre countless fish, turtles and cetaceans "for free"...

- Bottom trawling, whose simple concept consists of - literally- raking the bottom of the oceans, destroying everything in order to bring everything up, choosing what interests us and throwing the rest back into the water. It's a little as if in the forest, to hunt a species, we raze the whole forest, exterminating all the trees and other animals at the same time, to recover only the one we are interested in (a special thought to the particularly disturbed Homo sapiens, who was the first to have this idea).

- In order to extract its black gold from the tar sands, the oil industry does not hesitate to irreversibly destroy Canada's boreal forest. A primary forest with exceptional ecosystems that has taken millennia to develop...

- The dumping of waste -household or industrial- in the wild in Nature is practiced daily hundreds of millions of times, thus defiling hundreds of millions of places every day...

o Thousands of industrial accidents, large and small, with catastrophic consequences, occur every day all over the world...

o Pollution linked to nuclear accidents such as the one that occurred in 2011 in Fukushima, Japan, where the Pacific Ocean serves as a giant bin for colossal quantities of contaminated water and other radioactive waste...[42]

o The gold diggers of the Amazonian forests that pollute the rivers, destroy animal species and weaken ecosystems by massively dumping the mercury they use into them.

o All these millions of tons of unsorted and non-recycled waste buried in the ground... (A special thought for Amazon which, for reasons of space management in its warehouses and profitability, pushes vice to the point of massively burying brand new manufactured products)[43.]

o These tens of thousands of fireworks fired regularly all around the globe, spreading countless fine particles and traumatizing a multitude of birds...

o All the innumerable and varied pollutions linked to mass tourism...

o The recent emergence of all data centers and the massive pollution they generate...

...etc, etc, etc.

« Instead of assuming that the Earth belongs to us,
we must become aware that we belong to the Earth. »

Pierre Rabhi

- <u>Pollution of rivers</u>

Water covers 71% of the planet's surface. However, despite this apparent abundance of water on Earth, fresh water represents only 3% of the total quantity. And river water is even more scarce: it represents only 0.01% of all existing freshwater.[44] And this freshwater, which is believed to be inexhaustible and of which we all have a vital need - people, animals, trees, plants, fungi and bacteria - is under threat from all sides.

Our modern way of life consumes so much water that we even divert some of our rivers. More and more rivers are drying up and, whether large or small, massive pollution from multiple sources is poisoning river after river all over the world.

- <u>The looming water shortage</u>

Almost all the water on Earth is either trapped in ice or is salt water from the oceans. In fact, the freshwater available for the almost 8 billion people and an infinite number of living species corresponds to barely 0.02% of the total volume of water on the planet. No more than that. This water is at the center of the biosphere, the only part of our planet where life is possible. A set of ecosystems in which water, of course, but also oxygen, nitrogen, carbon and many other elements essential to life, are constantly being recycled.

And we use a lot of water. Much too much, because our personal use does not stop at our shower, our domestic needs and what we drink daily. Including everything we consume, as well as the water we use to produce and clean -what we call "virtual water"- the average Western man uses about 5 000 liters per person per day. If you want to see some examples of this virtual water, you should know that you have to:

- o 3 liters of water to produce a 1,5 liter bottle of mineral water

- o 40 liters to grow a salad

- o 140 liters for a single cup of coffee

- o 185 litres for 1 kilo of tomatoes

- o 330 liters for a single loaf of bread

- o 340 liters for 1 kilo of oranges

- o 960 liters for a bottle of wine

- o 1 000 liters for one kilo of apples

- o 1 100 liters for a liter of milk.

- o 1 900 liters for one kilo of pasta

- o 1 400 liters per kilo of rice

- o 2500 liters for a cotton shirt

- o 11 000 liters to produce one pair of jeans

- o   13 500 liters for 1 kilo of the beef you eat! (Because in addition to quenching its thirst, a very large quantity of water was used, among other things, to produce the cereals that the beef you ate was fed throughout its growth).

Our food, our lifestyles, our travel, our leisure activities and, of course, the population explosion: everything contributes to constantly increasing our water needs. And this need is growing so fast that it far exceeds what Nature can offer us. In order to meet our water needs, mankind is diverting rivers, drying up lakes and even using very old and non-renewable underground water resources. Added to this is the galloping urbanization and industries that increasingly pollute our reserves. Currently, 85% of the world's wastewater is untreated and is thus discharged directly into the environment. Not only is this water lost, but it also contaminates our available freshwater resources. We are poisoning ourselves!

« As a symbol of the emergence of technology and the scarcity of our resources, when I was a child, water was free and music had to be paid for. Now it's the other way around! (And that sucks!). »

Gérald Vignaud

Already today, in many parts of the world, water has become a scarce resource that is sorely lacking for more than a billion people. A deadly threat to some of the most vulnerable populations. At the beginning of the 21st century, 800 million people do not drink drinking water and contaminated water kills more than 4 000 children under the age of 5 every day.[45]

Please take a moment to reread and understand the real meaning of this last sentence: At the beginning of the 21st century, 800 million people do not drink drinking water and **contaminated water kills more than 4,000 children under the age of 5 every day**.

Freshwater is much rarer than we think: Because it covers nearly 71% of our planet, it might seem that water is its main component. But if on a human scale the oceans can be very deep, on a global scale they are absolutely not. The deepest point of the oceans is in the Pacific Ocean: the Mariana Trench, which is 10 984 meters deep and has a diameter of around 12 750 000 meters. In the end, if all the water on Earth - the water contained in the seas and oceans, lakes, rivers, groundwater and ice - were to be brought together, this colossal quantity of water would still only form a sphere 1 200 kilometers in diameter. This is the big blue ball in the picture opposite. The small blue ball next to it corresponds to the proportion of freshwater present on Earth. And the third ball, the tiny one that you can barely see, corresponds to the freshwater available - that means, not trapped in the ice- for all living things on the whole planet...

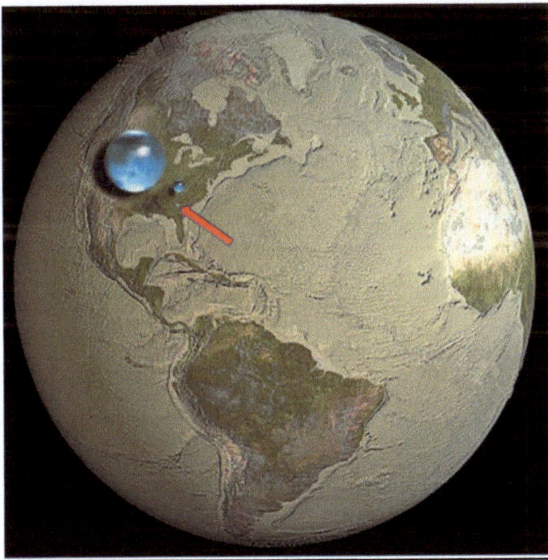

- <u>Shist gas</u>

Shist gas is natural gas found in large areas of our continent. As the name suggests, shale gas is contained in a rock, shale, which is found at average depths of 1 500 to 5 000 meters. The problem with shale gas is not the gas itself but the method used to extract it: hydraulic fracking.

Because shale has very low permeability, the gas it contains does not flow into reservoirs as it does with other types of rock. Hydraulic fracturing therefore involves injecting water at high pressure into the well. This creates fractures in the rock and allows the gas to flow. The water used is mixed with sand and a cocktail of a multitude of chemicals that makes up to 2% of the contents. Chemicals that lubricate and thus facilitate the flow of gas.

Approximately 50% of this highly polluting water can be recovered for treatment or reused for further hydraulic fracturing. **50% is not.** Concerning the first half of this highly toxic waste water, it is, in theory, recovered and recycled by the companies exploiting the deposits. Industrialists have, generally speaking, unfortunately got us used to optimizing their costs to the detriment of the environment. There is therefore a strong concern about the final destination of this water. And concerning the other half, the one that is not recovered, this poses a real problem. This chemical-laden water contaminates the soil and the water tables, thus multiplying the dangers for the environment, the ecosystems and the local populations.[46]

In our world where energy is a major issue, the revolution linked to the exploitation of shale gas in the United States has awakened interest in its potential in other countries of the world, particularly in Europe.

- <u>The problem of nuclear waste</u>

Nuclear power, professionals in the sector tell us, solves many problems. It covers our growing electricity needs, counterbalances the increasing scarcity of fossil fuels and provides for the indispensable need today to use energies that do not accentuate global warming. But the problems of nuclear power are major. First of all, the safety of the power plants - as we have seen with the accidents at Three Mile Island, Chernobyl and Fukushima - and their dismantling are today very complicated, if not impossible. But there is also and above all the problem of nuclear waste.

To date, it is not known how to recycle them and the solution of sending them into space has been abandoned because of the risks associated with a possible explosion of the rocket on take-off. So, we store them, bury them or throw them into the sea. In this regard, according to the International Atomic Energy Agency, in the second half of the 20th century, nuclear countries dumped more than 100 000 tons of nuclear waste into the oceans. But whether stored, buried or dumped at sea, some of this waste will take hundreds of thousands of years to completely lose its radioactivity. So, we are leaving this pollution to thousands of generations to come...

- <u>Pollution of space</u>

Since Sputnik and the beginning of the space conquest in 1957, nearly 8000 spacecraft have been launched by different nations. A great deal of debris has been dumped into space. They have accumulated almost exponentially and today it is a large mantle of a "high-tech" bin that rotates at very high speed in orbit around the Earth. Under the impact of the violent shocks linked to speed, these objects of all sizes subdivide into smaller and smaller objects, thus increasing the risk of collision. And all this is without counting the 42 000 (42 000!!!) additional satellites of the "Starlink" project that the billionaire Elon Musk would like to launch and position around the globe...[47]

Even if this pollution is much less serious and urgent than the many others that can unfortunately be found on Earth, the symbolism is nevertheless cruel. Even in space, Man has succeeded in the feat of polluting his environment.

So here is a quick summary of the major ecological challenges facing humanity at the beginning of the 21st century. Of course, among all these ecological challenges, some are bigger, more important, more complex and more urgent than others. Nevertheless, in addition to being all directly or indirectly interconnected, they all share the same cause: the development of human activities. They are the consequences of our current economic system based on maximum and infinite growth and the generalized dumbing down of populations. A crowd of people whose brains are very often poisoned by social networks and mass media.

« When they have cut down the last tree, polluted the last stream, caught the last fish, then they will realize that money cannot be eaten.»

Sitting Bull

Taking action, individually and collectively

In the millennia since the advent of Homo sapiens, our relationship with Nature has changed dramatically. From being animistic and mystical - in which we communicated directly and respectfully with each of the beings sharing the Earth with us - our relationship with our planet has gradually been transformed over time into a relationship of domination and economic exploitation, immoral and destructive.

Stuck by the mass media and imprisoned in their daily problems, the vast majority of people do not realize that our self-destruction has indeed begun. Today, everything is almost in place to finalize our own extinction and we don't have much time left to react. In the words of Paul Watson, if we want to survive on this planet, we will have to integrate that we have to live in accordance with the three laws of ecology:

- o The first law of ecology is **the law of diversity**: The strength of any natural ecosystem is based on its biodiversity. The more biodiversity it contains, the stronger an ecosystem is.

- o The second law of ecology is **the law of interdependence**: All species in an ecosystem are interdependent and need each other.

- o The third law of ecology is **the law of finite resources**: There is a limit to growth and a limit to the regenerative capacity of Nature.

Let's be clear: this is not advice, but truly laws of Nature that we must absolutely understand and respect. This respect is indispensable if we do not want to suffer the dramatic and lasting consequences of an imbalance in our ecosystems. But here we are caught in a destructive whirlwind that we have created ourselves. And the conclusion, if one has the courage to make an objective and factual analysis of the situation, is bitter. Like a cancerous cell, it seems that humanity has truly declared war on its entire planet and its ecosystems.

In our world in the midst of an identity, ideological and technological revolution, everything is moving very fast. Too fast. Whether ecological or human, the challenges we face are both unprecedented and colossal. Yet, faced with the immensity of what lies ahead, too few people seem to be able to truly understand the situation and respond effectively.

Prisoners of their beliefs and entangled in their personal problems, the overwhelming majority of people - those who form public opinion - are dazed in front of their TV and smartphone screens, without even perceiving the changes and perils that threaten us.

Without any global vision of things, the "experts" in all fields -when they are not at the service of the lobbies and multinationals- are often autistic and try to master only their fields of competence in which they are sometimes even completely lost in front of the too fast evolution of our world.

Not to mention the decision-makers -political and economic- who most of the time see and decide only through an ultra-liberal ideological prism and in their own short-term interests.

« You are undone if you once forget that the fruits of the earth belong to us all, and the earth itself to nobody. »

Jean-Jacques Rousseau

Protecting and preserving our planet will undoubtedly be the greatest, most difficult and most important struggle of all time. And it is happening now, at the beginning of the 21st century. Faced with this predicted ecological collapse, it has become more than urgent for us to react!

<u>First of all, we must react **individually, in** particular by:</u>

- o Our personal decision to really open our eyes to the current, factual reality of our world and the dangers that threaten it.

- o Proactive self-education to acquire a better personal knowledge of the functioning of our ecosystems, to know the causes of their current upheavals and to understand all their consequences.

- o Reducing our energy consumption wherever possible.

- o   Changing our daily actions and habits in terms of lifestyle and consumption.

- o   To support by our votes the few politicians who truly understand these ecological issues and who commit themselves to do everything possible to limit their damage as much as possible.

- o   A permanent reinvention of our way of thinking to strive for more personal inner growth rather than more material growth. More personal fulfilment rather than more money. More compassion rather than more self-centeredness. More long-term global vision, rather than more short-term personal vision.

- o   Contribute to the advent of a new vision of our world based on balance and respect for its resources and living beings.

- o   And why not, since it is better when life has meaning, a very serious evaluation of the possibility of dedicating one's life to helping solve one of the major challenges of the 21st century?

Pierre Rabhi often tells the story of the little hummingbird which, when its forest is burned, keeps going back and forth between the lake and the main fireplace, its tiny beak filled with water, to try to put it out. At one point, he comes across a majestic toucan with a large multicolored beak. Amazed by its attitude, the toucan pointed out to the little hummingbird that with its tiny beak what it is doing is almost useless. « I'm doing my part! » replied the little hummingbird...

Acting individually, each at his own level, will undoubtedly not be enough to resolve all these major challenges. However, it is essential for each of us, if only for moral reasons, to do "our part".

« When a piece of legislation contradicts the laws of Nature,
it is a moral duty to oppose it. »

Vandana Shiva

<u>We must then react **collectively, in** particular by:</u>

- A massive surge in world opinion on the dangerousness, growth and imminence of all these major challenges we face. In this respect, a voluntary re-qualification of the priorities given by the major media to their news would be welcome. Because no, a terrorist who blows himself up with a bomb in the metro, a thousand-year-old cathedral burning down in the center of Paris or the final of a football World Cup is in fact no more important - on the contrary, by far - than climate change, deforestation or the destruction of biodiversity.

- The management of all these challenges must be equal to what they should be: **urgent and major planetary crisis situations** at least comparable to those experienced during the Second World War or more recently during the global spread of COVID 19. This includes, among other things, very unpleasant decisions to be taken, the colossal financial releases needed and the mobilization of all.

- An ambitious international plan, signed and held by all countries for the protection of the oceans with the necessary financial, scientific, military and legal means.

o An ambitious international plan, signed and held by all countries for the protection of primary forests with the establishment of the necessary financial, scientific, military and legal means.

o An ambitious international plan, signed and held by all countries for the protection and preservation of freshwater reserves -lakes, rivers and groundwater- with the necessary financial, scientific, military and legal means.

o An ambitious international plan, signed and held by all countries, to completely clean up our planet in order to remove and recycle all the shit that is lying around and polluting our ecosystems.

o An ambitious international plan signed and held by all countries for waste management and recycling. (For your information, at the time of writing, rich countries are "selling" their waste of all kinds to poor countries who are supposed to recycle it for us...).

o To tax very heavily - or even prohibit - the production of waste that Nature cannot recycle on a human time scale. A very simple example with plastic: if it floods our planet - when biodegradable alternatives exist - it is simply because it is very cheap to produce. To slow down and then stop its production, all you have to do is tax it very heavily and invest all the proceeds in subsidies for the production of the alternatives.

o A total and global war against eco-predators of all kinds who do not hesitate to smash our planet to maximize the profits of their companies.

o International legal bodies that are genuinely empowered to enforce environmental protection laws already in force, as well as others, much stricter, that need to be urgently created.

- o Better management of our production and energy consumption.

- o A revolution which, more than political, economic or energy-related, must also and above all be spiritual.

More than an ecological revolution, a spiritual revolution

Even if it abounds in opportunities, our world is also, paradoxically, in great peril. Our planet is undergoing an unprecedented, gigantic transformation. More than ever, it has become essential for us to see ourselves not in a static way and at a given moment, but in movement through a global perspective. If we want to understand the current situation, we need to know where we have come from, where we are today and, above all, what are the fundamental reasons for these changes.

Man is the only species on Earth to ask for more and more. Today, the continuous growth in consumption by an ever-increasing world population is leading us inexorably to a short-term depletion of natural resources.

In order to make money, we methodically destroy, one after the other, almost every ecosystem on our planet. According to the *WWF's Living Planet 2018 report*[48], "Between 1970 and 2014, vertebrate populations - fishes, birds, mammals, amphibians and reptiles- fell by 60% globally and by 89% in the tropics and Central America."! The absolutely dramatic situation that we have produced is so overwhelming that you can no longer even realize the sentences that are used to describe it. To such an extent that these words become, for most of us, of great banality and pass through us without us even really understanding them.

So, I suggest that you reread this sentence: **"Between 1970 and 2014, vertebrate populations -fishes, birds, mammals, amphibians and reptiles- fell by 60% worldwide and by 89% in the tropics and Central America."** But this time, take a few minutes, now, to really think about it, to integrate its real meaning and all its consequences:

- o What does it really mean: "Between 1970 and 2014, vertebrate populations -fishes, birds, mammals, amphibians and reptiles- have fallen by 60% worldwide and by 89% in the tropics and Central America."?

- o How did it come to this?

- o What collective responsibility do we all have in this situation?

- o What is my individual responsibility, in terms of how I think, live and consume in this situation?

- o What are the consequences for our planet today?

- o Is this situation tending to get better or worse?

- o Does it empire in a slow and linear way or on the contrary in a fast and exponential way?

- How much longer can this go on?

- How much longer is this going to last?

- What impact will this have on our lives?

- What impact will this have on our children's lives?

- What will be the state of life on Earth in 20, 50 and 100 years (i.e. tomorrow)?

- What is our moral responsibility in all this?

- What can we do to change all this before it's too late?

The organization of the world as we Homo sapiens apply it at the beginning of the 21st century will structurally not last long. It is clear that a liberal economy that advocates exponential and infinite growth in a finite world is just not mathematically possible.

To change things and save our planet from the looming ecological collapse, it has become imperative to make a revolution. The systems that humanity has put in place over time are so complex, intertwined with each other and locked by such powerful interests that a revolution, be it political, economic or energy, seems unlikely. To get us out of this situation, **the only possible revolution, the one we must lead, is a spiritual revolution**. When I use the word spiritual here, I am not referring to religion, whatever it may be, but rather to something much more universal. I am talking here about that intuition which is omnipresent in each of us and which Teilhard de Chardin captured very well: We are not human beings living a spiritual experience, we are spiritual beings living a human experience".

« We are not human beings having a spiritual experience;
we are spiritual beings having a human experience. »

Teilhard de Chardin

We have to admit that we are only here for a spark of time on a tiny planet lost in the vastness of the Universe.

We must realize that this planet does not belong to us and that we are only invited.

We have to assimilate that on the scale of the Universe, we are all, without exception, from the President of the United States to the last of the drug addicts, micro-dust of shit, and therefore **we all have to put our egos on "off"**.

We must realize that any average Homo sapiens living in the 21st century has an ultra-positive ecological balance in general and carbon balance in particular (especially if he is Western). And more often than not, while contributing little or nothing to the positive evolution of humanity, he also takes the luxury of thinking that there is a "right" by brandishing the concept of "punitive ecology" as soon as he is asked to reduce his consumption, waste and more generally his impact on the resources of our planet. As a reminder, from his birth to his death, Mozart's carbon and ecological balance sheet is very close to zero, while the legacy he bequeaths to humanity borders on the infinite...

We need to understand that each of our actions that produce irreversible pollution -such as consuming plastic- is an ecological debt that is transferred to future generations. A debt that they will have the vital imperative to pay and to which they will have to pay colossal interest.

We need to challenge many, many individual and collective beliefs we have about ourselves and the world.

We must abandon the idea that only science and technology will liberate mankind.

We must realize that exponential and infinite growth -be it demographic, economic or technological- is neither natural, nor indispensable, nor preferable to balance.

Before it is too late, we must admit that the world, as we have made it evolve, can no longer hold together structurally and that we are heading straight for a wall.

Rather than accepting the things we cannot change, we must choose to change the things we cannot accept.

We need to revolutionize the way we act and also, and above all, the way we think.

We have to realize that between an industrial tomato produced in the South of Spain and sold at Carrefour for 1€ per kilo and an organic tomato produced close to home and bought in a Bio Coop at 3.5€ per kilo, the cheapest is not the one we think. Indeed, if we add the cost of the health impact (the cancer treatment you will have in 15 years time because of the pesticides in your tomato) and the ecological impact (the destruction of arable land and biodiversity because of these same pesticides as well as the contribution to global warming because of the fossil fuels used in its transport by truck across Europe), in the long term, the industrial tomato -despite its insipid taste and low nutritional value- costs much more.

We must understand that, directly or indirectly, everything is interconnected. To understand and resolve a situation in the most effective way, we must therefore develop a transdisciplinary approach to things.

We need to disconnect from screens and digital to reconnect with

Nature and ourselves.

We need to realize that there is far more intelligence and technology in any 'insignificant' plant or insect than in the latest smartphone. We therefore need to see Nature as a library from which we could benefit enormously by reading the books it contains rather than burning them one by one like a common fuel to heat us during the winter.

We must learn to see beyond what our eyes see. When you put an acorn in the palm of your hand, you can see just an acorn. But you can also open your eyes and your mind more widely to see that there are billions of years of evolution behind it that have recently resulted in a magnificent line of majestic oaks, the last of which gave birth to this hopeful little acorn. A little acorn that has the potential to become a shoot, then a tree, then a forest...

We must initiate ourselves into meditation and listening to our connection with ourselves, with Nature and with the Universe.

And we must share this philosophy with our children by initiating them too to meditation and listening to their connection with themselves, with Nature and with the Universe.

We must give special care to the education we give to our children by giving them time, kindness and love. Lots of love! We must teach them respect for themselves, others, animals and the planet.

And since education is the most powerful weapon we can use to change the world, we must democratize Maria Montessori's pedagogy. Since it is unanimously recognized as one of the most effective of all because of its results, we must make it the default standard used in all our public schools and education systems.

We must transform our market economies (based on growth and over-consumption) into a resource economy (based on the balance and preservation of our resources and ecosystems).

We must save the last great mammals before it is too late.

We must deplastify our civilizations and cleanse our planet of all the waste that disfigures it.

We must detoxify ourselves from fossil fuels and stop global warming in order to limit its consequences.

We urgently need to regenerate our soils.[49]

We must immediately stop flooding all the ecosystems of our planet with all kinds of chemicals.

We must fight for what we believe is right.

We must respect ourselves better and reinvent the way we eat by refusing to consume all those things that destroy our bodies and, very often, the planet: junk food, intensive farming and agriculture, GMOs, industrial food, food chemistry, etc. etc.

We need to reinvent our relationship with medicine, which today has almost no connection with health and well-being.

We need to get rid of the "consumer God", transform our relationship to material things and lighten our lives.

We must imperatively protect and preserve the Commons: water, air, soil, seeds, climate.[50] All these things are neither private nor public goods but, as the name suggests, common goods that are indispensable to the life of all, including plants and other animal species.

We must understand and accept that every animal on this planet, from the grasshopper to the polar bear, the vulture, the mackerel or the spider, has **the same legitimacy to live here as we do**. Consequently, we must assume our duty to protect them individually as beings and to preserve them collectively as a species.

We must preserve our ecosystems with special care for our forests, swamps, lakes, rivers, mangroves, coral reefs, seas and oceans.

We must not only defend Nature, but we must go beyond it by reconnecting with it to become a part of it again.

We must assume our moral responsibility to protect our planet and pass it on to those to come, regardless of the generation or species to which they belong.

« Planet Earth is a beautiful and fragile object that is home to a mixture of life forms that are certainly rare, if not unique, in the entire Universe and the preservation of its integrity is now the privilege and responsibility of humanity. »

Ervin László

We must integrate that the secret of life is giving!

We need to understand that the Universe is a holistic structure. Therefore, we are at the same time a whole made up of a multitude of smaller elements as well as one of the tiny elements of a much larger whole than us, itself a tiny element of an even larger whole and so on...

We must understand that the Universe is a colossal force that is beyond us and we must trust it.

Our only main focus should be on: Loving, Growing and Giving.

Individually and collectively, we need to access higher levels of consciousness.

All of us, whoever we are, must become aware of the global problem facing our planet and commit ourselves to help solve it. And "all of us, whoever we are", starts with you. Ask yourself:

What does the word "ecology" really mean to me?

____

____

____

____

____

____

____

____

____

____

____

____

____

____

What is my level of knowledge and understanding of current ecological challenges?

____

____

____

____

____

____

____

____

____

Are these topics of interest to me? Why?

____

____

____

____

____

____

____

____

____

____

What is my point of view on the current ecological crisis? Why?

Am I aware that we are running in the very short term towards a major ecological collapse?

☐ Yes   ☐ No

If my answer is no, isn't it wise -given the importance of the subject- that I sincerely take the time to study the question?

☐ Yes   ☐ No

If you have answered *Yes* to this question, I recommend the book by Paul Watson and Lamya Essemlali: "Interview with a pirate". If you answered *No*, I can only advise you twice as much... (and if you really don't like reading, type their names on YouTube and take the time to listen to one or two of their videos on this subject).

« You are not born as an ecologist, you become one. »

Nicolas Hulot

What are the individual behaviors I can decide to adopt in my daily life to limit my negative impact on the planet (e.g., to reduce the impact of the climate change on my family, my friends, my family's environment, etc.? Stop buying useless gadgets, Repair rather than replace, Limit -or even eliminate- my consumption of meat, Replace the plane by the train, the car by the bicycle, Limit my energy consumption whenever possible, Support financially clean and committed NGOs such as *Sea Shepherd, Rewild, One Voice, SOS Ocean*, the *GoodPlanet* foundation or the *WWF*...etc.)

____

____

____

____

____

____

____

____

____

____

____

____

____

____

____

____

____

____

____

____

How can I contribute to impacting collective behaviors that go in the direction of solutions (e.g. I vote for people who understand these problems and who are really committed to solving them, I denounce and participate in putting pressure on eco-predators of all kinds, By my attitude, I value positive behaviors in order to positively influence social conformism, I contribute, by example, to the advent of a higher level of awareness...).

Of all the major ecological challenges of the 21st century, which ones particularly upset me? Why?

____

____

____

____

____

____

____

____

____

____

____

If I had to choose just one to act on, which one would it be? Why?

____

____

____

____

____

____

____

____

____

____

____

And since it is better when life has meaning, would I be willing to seriously consider dedicating my life to helping solve one of the major challenges of the 21st century? Why?

If so, what are the first steps I can start to take today? (e.g. I find out more about the issue by questioning the Internet, I order books on the subject, I read "EARTHFORCE - An earth warrior's guide to strategy" by Captain Paul Watson, I meet people who have mastered the subject, I join associations that work on it, I create a foundation and set up projects...).

« You never change things by fighting the existing reality. To change something, build a new model that makes the existing model obsolete.»

Richard Buckminster Fuller

What kind of world do we want to leave to our children?

The 21st century is dawning on a sick planet Earth where more and more people are consuming -and wasting- more and more resources as they become poorer and less quickly renewed. If humanity does not take **a radical turn** in its way of thinking and functioning, we are heading straight for a chain reaction of major ecological cataclysms with unpredictable consequences!

I believe that the greatest danger to our planet is the common belief that someone else will save it. It is now imperative for us to hear the crucial question that this beginning of the 21st century poses to each and every one of us, individually and intimately.

More than ever, the world needs real and committed leaders.

Will you answer the call?

Conclusion*

---

*This is the conclusion of the book "School is important but education is paramount !" from which is extracted the book you are holding in your hands. I added it in conclusion of all the 5 books of the eponymous collection from which it comes from (see page 131). I am specifying it to you because, without explanation, some of the elements of this conclusion could, rightly, seem irrelevant to you.

« Knowledge is a weapon, now you know it! »

Stomy Bugsy

Throughout their lives they had worked hard and never taken a single day off. According to the most common social criteria in our society, this trading couple had succeeded brilliantly. After almost 35 years of hard work, they were at the head of about fifteen quality bakeries, all located in the best areas of the capital. They had about a hundred loyal employees and their business was going very well. And this was no coincidence: for 35 years they had been managing their money very carefully, not making any unnecessary expenditures and reinvesting everything they could. They had been doing this for so long that saving had become second nature to them, to the point that they had never gone on holiday together.

One evening, during the meal, after a long and tiring working day, the woman had a chance to come up with an idea:

- Tell me, darling, we've never been on holiday. What would you say if we took a cruise? I saw that there are promotions at the moment...

Her husband was torn. Deep down he wanted to leave, but he had created a powerful programme for himself that required him never to spend money unnecessarily. After a few seconds of reflection, he proposed to his wife:

- Why not, it will allow us to disconnect, we've earned it, after all. But on the other hand, we will be very careful about our expenses on the spot.

A few days later, they embarked for a week on one of the jewels of the Cunard company: the splendid Queen Mary II. In order to stay within the correct budgets, they had planned everything to keep costs as low as possible and, in order not to have to pay for the restaurant, they packed sandwiches in their suitcases.

So, the journey began and went superbly well. They took time for themselves, lounging around the pool with a book in their hand, and took part in all the free activities during which they met many people with whom they sympathized. But every time it was time to eat, it was the same scene: the wife told their travelling companions that she wasn't feeling well and, together with her husband, they went to their little cabin without a window where they ate their provisions.

Their week of holidays went like this and, apart from during meals, they had a great time. On the last evening of the cruise, when their provisions were almost exhausted and their last piece of bread had been staling for four days already, the woman proposed:

- Honey, we probably won't be going on holiday for a long time. Instead of eating more stale bread and cheese tonight, what do you think about having a restaurant together for our last evening on the boat?

Delighted by the idea, her husband agreed.

When they arrived at the restaurant, the waiter offered them a table with an absolutely sumptuous view of the setting sun. The background music was soft and the evening looked like a pleasant one. When the waiter handed them the menus, they were surprised not to see any prices associated with the dishes. When ordering, the husband then asked how much the one he wanted to choose cost. Surprised, the waiter simply told them that he could take whatever they wanted because the meals were included in the price of their trip...

« Life is too short to be small. »

Tim Ferriss

Life is a gift!

I believe that life is a gift! A journey that has been given to us and that we have the opportunity to enjoy to the fullest. Of course, when I talk about enjoying the trip to the fullest, I am not thinking here of material things, but of experiences, accomplishments, personal development, contributions and the quality of emotions we are given to experience. Most of us are conditioned to accept that we live far below our potential and possibilities for fulfilment. And unfortunately, it is often only at the end of the journey that we realize this and bitterly regret some of our choices.

It seems that the definition of hell is **when, at your death, the person you are meets the person you could have become**. Ask yourself: if you die today, will you go to hell?

If the answer to this question is yes, which is statistically more than likely, it may be time for you to react. Regardless of your age and where your life is today, decide to take control of it again. It's not necessarily simple, but I have some great news for you: it's not what you've done so far that counts, it's really what you do from now on.

« It's not what you've done so far that counts,
but what you're going to do from now on. »

Gérald Vignaud

But to counterbalance this excellent news, there is also something you absolutely must take on board: if you don't apply it, all learning is fruitless. Which, in short, means that if you don't take action after reading this book, then it will have been useless to you.

When you were a child, you could only endure and you were a victim of circumstances. But now that you are an adult, you are a victim of your decisions. No matter what the timing of your life, the best time to regain control of it is now! There are no more excuses, because you have the first keys to a successful life and the directions in which to go deeper in your quest. Understand this: today is the first day of the rest of your life! Take out that famous dream box we were talking about earlier. Open your mind, let it run without judgment and decide which ones you want to conquer. Reread the personal notes you took while reading this book and take some time to think about them. Who are you really? What are your past experiences, your successes, your failures, your emotions, your intuitions? Who do you really want to become? And most importantly, whether it is at the level of your community or the planet, how do you want to positively impact the world? Go back to the chapter about your goals and take action. Take action today!

Sometimes the days seem long to us but, in the end, time flies and the years are short. Never lose sight of the fact that in a few decades - a hundred years at the most - you will be dead and so will all your loved ones. In fact, in barely a century, almost all the current inhabitants of the planet will have renewed themselves. So, don't waste your time, because for you it's now. Tomorrow will be too late!

Change the world?

As Helen Keller so aptly put it, « Life is either a daring adventure or nothing! ». And since you only have one, decide to live a rich and exciting life. Be full of positive information, meet interesting people and be curious about everything: you will always find something exciting and constructive to do. Observe, listen and learn. Develop new ideas and explore new paths. And since the world is shaped by unreasonable people, why don't you decide to finally be yourself, even if it means being really unreasonable?

And besides, while we are talking about unreasonable projects. Ask yourself: do you really like the world we live in? If not, what would you like to change? And if you had the power in your life to radically change one thing, **one thing only**, what would it be? It's important that you really ask yourself this question, because the good news is that you already have the power to do so!

If there is one thing we unfortunately do not learn at school, it is that a determined person who acts with passion, vision, strategy and intelligence is capable of uniting a community of committed people and launching a real dynamic. A dynamic that, with time and cumulative effect, can transform the face of the world forever. If you decide it, that person can be you. And besides, if it is not you, then who will it be? And if not now, then when?

« Never doubt that a small group of thoughtful, committed, citizens can change the world. Indeed, it is the only thing that ever has. »

Margaret Mead

Whatever happens, never lose sight of the fact that one day your heart will stop beating. And on that day, none of your fears, none of your hesitations, none of your doubts, none of your regrets and none of your future goals will matter anymore. From then on, the only things that will really matter - and that will remain engraved for eternity - are how you lived, what you learned, what you understood, what you felt, what you gave, how you contributed, the love you shared and how far you managed to push your spiritual awakening.

The secret of life is giving

A final word to conclude. Personally, I believe deeply in the law of Karma, the law that says the more you give, the more you receive. A wisdom that I have been trying to integrate into my life for many years now.

I believe that giving anonymously and sincerely selflessly sends energy to the Universe. An energy that triggers in return an amplified boomerang effect that will penetrate your destiny. I truly believe that the ultimate secret of existence is to give and I would like to propose that you apply this philosophy to your life. If you find that this book has helped you in any way, consider giving a copy to five people you care about and think it might be useful. This could be family members, friends, co-workers, a business partner or even a casual acquaintance whose life you would like to make a positive difference.

When I was proofreading the final manuscript of this book, one of my friends advised me to remove the last paragraph, saying that readers would think it would be in my interest, which is factually true. After some thought, I thought that I should leave it anyway, as its consequences would also benefit many other people. On the one hand, it will have a positive impact on the lives of the people you decide to give the book to. But above all, this gesture will give you the priceless privilege of having added value to the lives of others. And perhaps even, in some cases, to have made a radical and positive difference.

Who are the five people to whom I am going to give a copy of this book?

- o 1 ____

- o 2 ____

- o 3 ____

o 4 ______________________________________

o 5 ______________________________________

And since it is now time to conclude this book, I would like to thank you for your confidence in me for having bought it and read it to the end. I wrote it with commitment and passion. I hope that it has been profitable for you and that it will contribute to your development. If you have enjoyed this book, do not hesitate to share it with others and to share your impressions on social networks. Don't hesitate either to leave a nice comment on the Amazon website (and/or other online distributors). This is very useful to me because, in addition to helping future readers choose the book, Amazon's algorithms estimate the popularity of a book by the number of comments left. And the more popular a book is, the more favorably Amazon positions it in searches.

Also, if you want us to continue the journey together, you can join me on my website (geraldvignaud.com) to discover other tools and the trainings I have conceptualized.

Finally, as I mentioned at the beginning of this book, I believe in the importance of horizontal communication. So do not hesitate to write me a message directly (www.geraldvignaud.com/livre-contact) to share your feedback and success stories after reading this book. I personally read all the messages and try to answer them as often as possible.

See you soon,

Friendly,

Gérald Vignaud

<u>Sources and further information</u>

[1] The film ''ZEITGEIST: Moving Forward'', the third part of Peter Joseph's exceptional trilogy.
Available on YouTube:
https://www.youtube.com/watch?v=ZoEVJQTUqdw

[2] on our planet, 160 000 people die every day for every 400 000 born.
Source: Documentary ''Terra'' by Yann Arthus-Bertrand available on YouTube: https://www.youtube.com/watch?v=nyIeWmkzEqs

[3] Taking previous centuries as a point of comparison, humanity in the 21st century seems to have really limited the ravages of wars, epidemics and famines, the three greatest causes of mass death in its history
Unless the evolution of the COVID 19 epidemic, which has just set the world on fire as I write this, makes me lie?

[4] As a reminder, today, 45 million people on Earth live a life of slavery. This estimate actually dates from 2016, but it seems that this figure has unfortunately not decreased since
https://worldpopulationreview.com/country-rankings/countries-that-still-have-slavery

[5] An epidemic of obesity that now affects more people on the planet than famines
According to the WHO, in 2016, more than 1.9 billion adults -people aged 18 and over- were overweight. Of these, more than 650 million were obese.
https://www.who.int/fr/news-room/fact-sheets/detail/obesity-and-overweight

[6] For example, a single French fry in a fast-food restaurant can contain up to ten different additives.
https://www.washingtonpost.com/news/wonk/wp/2015/01/22/there-are-19-ingredients-in-mcdonalds-french-fries

[6] On average, obesity leads to a loss of life expectancy of about 10 years.
https://www.sciencedaily.com/releases/2009/03/090319224823.htm

[7] In early 2016, a British report announced that by 2050, antibiotic resistance would cause the deaths of 10 million people worldwide each year if nothing changes.
https://www.who.int/news/item/29-04-2019-new-report-calls-for-urgent-action-to-avert-antimicrobial-resistance-crisis

[8] Numerous experiments, some of which have been carried out on rats (whose DNA is very close to that of humans) have shown that the waves promote the development of cancerous tumors and damage neurons...
https://www.sciencedirect.com/science/article/pii/S0386111121460069
8

[9] This poses a very serious health problem because, as you probably already know, sleep is a crucial time for rest and the formation of neurological connections. Something particularly important and active during the period of adolescence.
https://www.sleepfoundation.org/teens-and-sleep

[10] To take *Star Wars* as an example, countless fans know absolutely everything about this universe, including a multitude of improbable details such as the exact size of the Death Star, the raw material used to make Han Solo's blaster, the cruising speed of this or that ship, or the details of the childhood of this or that secondary character in the film.
If you wish, you can of course also find all this information on the many fan sites, such as this one:
https://starwars.fandom.com/wiki/Star_Wars:_Absolutely_Everything_You_Need_to_Know

[11] isn't it hallucinating to discover that more than 9% of French people believe "it is possible that the Earth is flat and not round as we have been told since school"?
https://www.nationalgeographic.fr/sciences/un-francais-sur-10-pense-que-la-terre-est-plate

[12] one quarter of Americans think that it is the Sun that revolves around the Earth
https://abcnews.go.com/US/quarter-americans-convinced-sun-revolves-earth-survey-finds/story?id=22542847

[13] 7% of American adults (which still makes more than 16 million) believe that chocolate milk comes from brown cows?!?
https://www.washingtonpost.com/news/wonk/wp/2017/06/15/seven-percent-of-americans-think-chocolate-milk-comes-from-brown-cows-and-thats-not-even-the-scary-part/

[14] More info on "filter bubbles":
https://en.wikipedia.org/wiki/Filter_bubble

[15] But, as Edward Snowden rightly points out, the very essence of being and personality of every human being cannot be built without privacy and intimacy.
https://www.different.land/sinspirer/reflexions/limportance-de-vie-privee-edward-snowden.php

[16] Today, a hotel in Nagasaki, Japan, welcomes its guests exclusively with the first generations of intelligent robots.
https://www.theguardian.com/travel/2015/aug/14/japan-henn-na-hotel-staffed-by-robots

[17] Meanwhile, an artificial intelligence made by Google is winning against Lee Se-Dol, the world champion of the game of Go.
https://www.theatlantic.com/technology/archive/2016/03/the-invisible-opponent/475611

[18] Following the "Human Genome Project" programme undertaken in 1988, the 21st century opened with a major scientific event: the sequencing and assembly of the human genome.
https://en.wikipedia.org/wiki/Human_Genome_Project

[19] "cows with porthole".
https://www.bbc.com/news/world-europe-48718908

[20] as Jean de la Fontaine's fable "The man and the adder" testifies
https://www.lafontaine.net/lesFables/fableEtr.php?id=901

[21] 40 quadrillion
To find out which numbers come after "billion":
https://blog.prepscholar.com/what-comes-after-trillion

[22] From an objective analysis of these empirical data, one deduction is obvious: the possibility that we are the center of a Universe that was created exclusively for us, Homo sapiens, is statistically absurd. The probability is so infinitesimal that it is most likely equal to zero.
To find out more, take the time to watch the series "Une espèce à part" produced by Arté. Beware, this series is absolutely exceptional!
It is available on YouTube:
https://www.youtube.com/watch?v=stCxLxBMjYA

[23] Over the last 7 million years, several species of hominids have succeeded one another.
https://www.nature.com/scitable/knowledge/library/overview-of-hominin-evolution-89010983

[24] such as a meteorite ten kilometers in diameter that strikes the Earth off the Yucatan Peninsula in present-day Mexico.
https://en.wikipedia.org/wiki/Chicxulub_crater

[25] While humanity consumes more than 95 million barrels of oil every day
https://en.wikipedia.org/wiki/List_of_countries_by_oil_consumption

[26] Like Peter Diamandis, they dismiss the dangers of today's ecological challenges out of hand, explaining that there is no need to worry, that tomorrow's technological discoveries -without specifying which ones- will solve them all.
Source: TED by Peter Diamondis, at 2.18:
https://www.youtube.com/watch?v=BltRufe5kkl

[27] Instead of wanting to Terraform the planet Mars
Like Elon Musk, many would dream of being able to artificially create an atmosphere on Mars in order to colonize it. They call this idea "terraformation":
https://thehill.com/opinion/technology/381262-could-mars-ever-be-a-new-home-for-humans

[28] wouldn't it be much wiser to try not to Venusform the planet Earth?
On Venus, due to a colossal greenhouse effect, the atmosphere rises to 490°.
https://courses.lumenlearning.com/astronomy/chapter/the-massive-atmosphere-of-venus

[29] At the Madrid' COP 25, in 2019, the idea took on even greater significance when some people publicly questioned whether it was wise to (attempt to) manipulate the oceans through geo-engineering.
https://www.lexpress.fr/actualite/sciences/cop-25-doit-on-manipuler-l-ocean-pour-sauver-le-climat_2109201.html

[30] In the last 40 years, we have lost almost half of the living and all the lights are red for a large part of the other half that remains.
https://livingplanet.panda.org

[31] the largest of which can carry more than 23 000 containers.
https://www.marineinsight.com/know-more/top-10-worlds-largest-container-ships-in-2019

[32] it is estimated that a single boat emits as much sulphur as 50 million cars!
https://www.greencarreports.com/news/1020063_pollution-perspective-one-giant-cargo-ship-emits-as-much-as-50-million-cars

[33] Displacement of invasive species caused by deballasting
https://www.sciencedirect.com/topics/agricultural-and-biological-sciences/ballast-water

[34] Unrecycled wrecks that end up on rubbish beaches in countries such as India or Bangladesh where the rest of their remains continue to pollute indefinitely...
https://shipbreakingplatform.org/our-work/the-problem

[35] the opening of new trade routes to the North made possible by global warming
https://www.bbc.com/news/business-45527531

[36] As sharks are top predators at the very top of the food chain, their extinction, through a domino effect, will unpredictably disrupt all our ecosystems.
https://www.mysticaquarium.org/2017/07/26/need-sharks-healthy-oceans-healthy-planet

[37] it is estimated that to produce just 1 kilo of beef, 13 500 liters of water is needed in the end
http://www.unesco.org/new/fileadmin/MULTIMEDIA/FIELD/Venice/pdf/special_events/bozza_scheda_DOW04_1.0.pdf

[38] the flatulence of these billions of cattle that release into the atmosphere phenomenal quantities of methane, a greenhouse gas 25 times more powerful than CO2.
https://letstalkscience.ca/educational-resources/stem-in-context/cows-methane-and-climate-change

[39] 1 gram of soil contains an average of 1 billion bacteria, divided into 10 to 100 000 different species, most of which are completely unknown to us.
https://www.novozymes.com/en/news/news-archive/2015/10/what-microbes-mean-for-agriculture

[40] In a country like France, however, it is estimated that at least 48 000 people die each year from air pollution.
https://www.thelocal.fr/20160621/pollution-in-france-kills-48000-people-each-year

[41] and often useless
such as the public lighting of certain deserted places at night, the signs of certain shops or even these huge company buildings illuminated all night long when there is nobody in them...

[42] Pollution linked to nuclear accidents such as the one that occurred in 2011 in Fukushima, Japan, where the Pacific Ocean serves as a giant bin for colossal quantities of contaminated water and other radioactive waste...
https://en.wikipedia.org/wiki/Fukushima_Daiichi_nuclear_disaster

[43] (A special thought for Amazon which, for reasons of space management in its warehouses and profitability, pushes vice to the point of massively burying brand new manufactured products).
https://www.retaildetail.eu/en/news/general/amazon-destroys-millions-unsold-products

[44] And river water is even scarcer: it represents only 0.01% of all existing freshwater.
https://www.usbr.gov/mp/arwec/water-facts-ww-water-sup.html

[45] Contaminated water kills more than 4 000 children under the age of 5 every day.
https://www.unicef.org/media/media_21423.html

[46] And concerning the other half, the one that is not recovered, this poses a real problem. This chemical-laden water contaminates the soil and the water tables, thus multiplying the dangers for the environment, the ecosystems and the local populations.
https://www.greenpeace.org/usa/global-warming/issues/fracking/environmental-impacts-water

[47] And all this is without counting the 42 000 (42 000!!!) additional satellites of the "Starlink" project that the billionaire Elon Musk would like to launch and position around the globe...
https://www.msn.com/fr-fr/actualite/technologie-et-sciences/starlink-pourquoi-il-faut-redouter-la-flotte-de-satellites-delon-musk/ar-BB13XNxV

[48] According to the *WWF's Living Planet 2018 report*
https://wwf.be/assets/RAPPORT-AUTRES/LPR2018-Full-Report.pdf

[49] We urgently need to regenerate our soils.
If this theme speaks to you, I'll let you discover "Blue Soil", the innovative concept of the young farmer-inventive-ecofeminist Céline Basset.
https://www.bluesoil.org

[50] We must imperatively protect and preserve the Commons: water, air, soil, seeds, climate.
https://www.different.land/construire/solutionner-grands-defis-21eme-siecle/solutionner-grands-defis-ecologiques/preserver-les-communs.php

---

Find here a summary and links of all the works recommended in this book:

https://www.geraldvignaud.com/liste-des-references

<u>**By the same author**</u>

GÉRALD VIGNAUD

SCHOOL IS IMPORTANT
BUT EDUCATION
IS PARAMOUNT !

—

The 15 essential things for success
that you will never learn in school

School is important
but education is paramount !

You got a problem ?

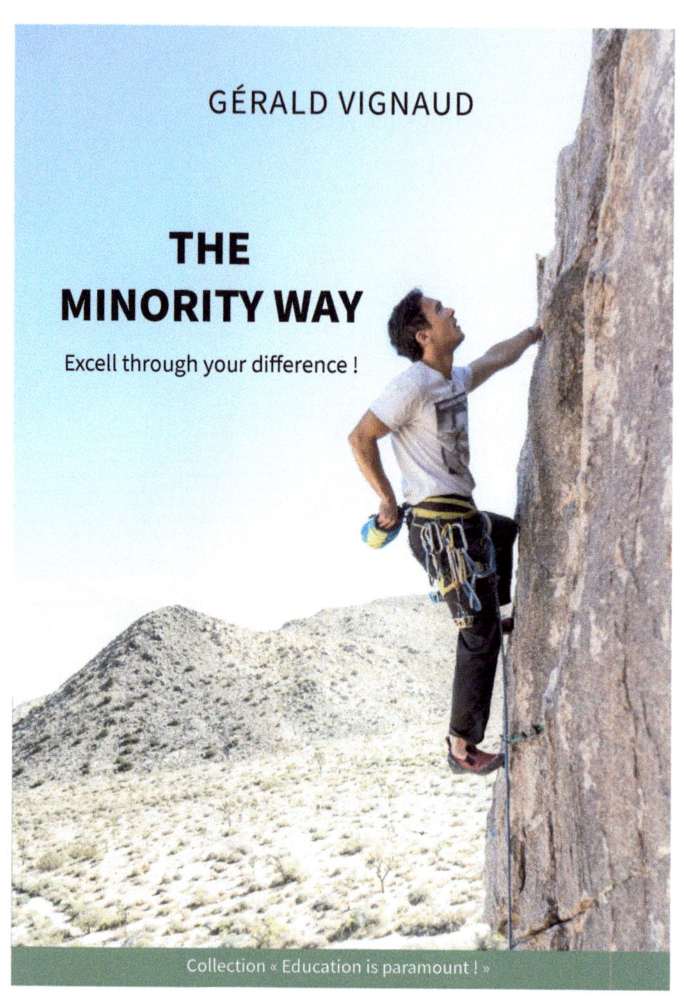

The minority way

GÉRALD VIGNAUD

BE HEALTHY !

The keys to protect your immune system
and maximize your energy

Collection « Education is paramount ! »

Be healthy !

Turn your dreams into reality !

Top performance !

About the author

Gérald Vignaud is a graduate of the *Business Mastery and the* prestigious *Mastery University. He* has had an exceptional career in network marketing, becoming *Senior Vice President of* one of the most important companies in the industry. As a recognized expert in the industry, Gérald has trained and coached tens of thousands of people.

As a personal development coach, he has always taught that the key to success and, more importantly, to personal fulfilment lies in accepting oneself by assuming and working on one's difference, one's X-factor.

As an expert in personal transformation, his first client was himself. As an addict for almost 10 years, Gérald was able to make decisions and take action. He has implemented the strategies he now teaches to get off drugs and propel his life towards exceptional personal and professional success.

For several years, he has inspired, advised and worked with many people from all social/professional categories, including self-employed people, business leaders, top athletes, politicians and celebrities.

As a business consultant, Gérald understands and possesses the keys and strategies to help companies in all sectors reinvent themselves. He helps them re-energize themselves to redirect them towards the more positive and stable results they desire.

A renowned speaker, Gérald regularly appears before audiences ranging from 100 to 15 000 people and has shared the stage with personalities such as Chris Widener, Darren Hardy and Donald J. Trump.

As an interviewer and passionate traveller, Gérald has listened to and learned from the thousands of people he has met over the course of his life.

As an iconoclastic entrepreneur, Gérald is the founder and CEO of different.land, a personal development platform unique in the world. It focuses on the idea that the three major keys to creating a better future are education, ecology and healthy technology, and that all three are interconnected. different.land website aims to help educate, inspire and nurture a new generation of leaders. One that will be responsible for building a much-needed better world for tomorrow, one that we will leave to our children.

As a true nature lover, Gérald is notably the co-founder of the association *SOS Ocean*. its mission is to inform, to make people understand the challenges that threaten our seas and oceans and to act to try to preserve them. More generally, Gérald also campaigns for better resources management and greater respect for our planet's ecosystems.

As a true "Learning Junkie", Gérald constantly seeks to learn, to reinvent himself and to set the bar higher and higher in his life.

Gérald's mission is to help build present and future generations by helping people develop their differences and multiply their personal, professional and financial values.

If you would like Gérald Vignaud to speak at your event,
please contact us directly via the website:

www.geraldvignaud.com

<u>A few words about SOS Ocean</u>

Our planet in general and our seas and oceans in particular are going badly. I believe that each of us, at his own level, has the responsibility to act. It is in this logic that with a few friends we have created the NGO *SOS Ocean*.

Its mission is threefold:

- To educate for a better knowledge of the ecosystems of our seas and oceans.

- To inform a maximum number of people of the dangers that threaten them.

- To act for their protection and preservation with a particular focus on issues related to plastics.

I donate all the benefits of this book to this structure. By purchasing a copy of this one, you have contributed to this important mission. If you want to know more and/or continue to support us, join us on www.sosocean.org and on social networks:

SOS Ocean is a member of the "1% for the planet" network.

https://www.onepercentfortheplanet.fr

Process of constant and perpetual improvement

This book is far from perfect. However, like all the things I try to do in my life, it follows a process of constant and perpetual improvement.

If you come across any typos, errors or other things that you think are inaccurate, don't hesitate to let me know so that I can make the necessary corrections in later versions.

www.geraldvignaud.com/livre-contact

"I alone cannot change the world; but I can throw
a stone on the waters to create many ripples. »

Mother Teresa